BIOMATERIALS
An Introduction

BIOMATERIALS
An Introduction

Joon Bu Park
Clemson University
Clemson, South Carolina

PLENUM PRESS · NEW YORK AND LONDON

Library of Congress Cataloging in Publication Data

Park, Joon Bu
  Biomaterials.
  Includes bibliographies and index.
  1. Biomedical materials. I. Title.
R857.M3P37   617'.95'028   78-12542
ISBN 0-306-40103-7

First Printing – April 1979
Second Printing – April 1980

©1979 Plenum Press, New York
A Division of Plenum Publishing Corporation
227 West 17th Street, New York, N.Y. 10011

All rights reserved

No part of this book may be reproduced, stored in a retrieval system, or transmitted,
in any form or by any means, electronic, mechanical, photocopying, microfilming,
recording, or otherwise, without written permission from the Publisher

Printed in the United States of America

PREFACE

This book is written for students who want a working knowledge in the field of implant materials. Obviously, the interdisciplinary nature of this subject has been a major obstacle in writing a book of this nature.

In writing this book, I have attempted to cover both biological and nonbiological (man-made) materials for obvious reasons. Hence, this book can be divided into three parts—man-made materials, biological materials, and implant materials.

The fundamental structure–property relationship is dealt with in the beginning, followed by the biological materials.

Implant materials or biomaterials as such are not greatly different from other man-made materials. Therefore, their acceptability in the body is emphasized. In addition, the reasons for a particular implant design and its material selection have been given special attention.

An effort is made to convert all the units into SI units although one or two exceptions are made such as Å ($= 10^{-10}$ m). Also some abbreviations such as v/o (volume %) and w/o (weight %) are used for brevity.

To cover the wide range of subjects dealt with in this book, I have used countless original and review articles as well as my own research proposals. A conscientious effort has been made to give credit to the original sources. Credit is given in the captions of the illustrations. For the occasional oversight of some tables and figures which could not be traced, the author offers his apologies.

It is recommended that the Further Reading sections and the references in tables and figures be consulted for an in-depth study on any of the subjects given. Also, the examples given in the text and problems at the end of each chapter should be solved by the student to increase familiarity with the current problems in the field of implants.

Finally, the author is greatly indebted to Dr. G. H. Kenner, whose contributions to this book are numerous, including his critical reading of the original manuscript.

Special mention should be given to Mrs. Regina Freeman, who skillfully typed the manuscript. Many valuable suggestions were made by Mr. Steven Young, Dr. Charles Fain, and Dr. James Wolf.

The constant encouragement given to me by Dr. Francis Cooke and my wife and children has been the major factor for finishing this book.

Joon Bu Park

Clemson, South Carolina

CONTENTS

Chapter 1 • INTRODUCTION 1
Further Reading 6

Chapter 2 • CHARACTERIZATION OF MATERIALS 7

2.1. Mechanical Properties 7
    2.1.1. Stress–Strain Behavior 7
    2.1.2. Mechanical Failure 12
    2.1.3. Viscoelasticity 13
    2.1.4. Pure Bending Stress 18
2.2. Thermal Properties 22
2.3. Surface Properties and Adhesion 24
Problems 26
Further Reading 28

Chapter 3 • THE STRUCTURE OF SOLIDS 29

3.1. Bonding between Atoms 29
3.2. Arrangement of Atoms 30
    3.2.1. Atoms of the Same Size 30
    3.2.2. Atoms of Different Size 33
3.3. Imperfections in Structures 34
3.4. Amorphous Solids 36
Problems 37
Further Reading 38

Chapter 4 • METALS AND ALLOYS 41

4.1. Alloys and Phase Diagrams 41
4.2. Imperfections and Strengthening Mechanisms 46

4.3. Corrosion of Metals	48
Problems	54
Further Reading	57

Chapter 5 • CERAMIC MATERIALS ... 59

5.1. Atomic Bonding and Arrangement	59
5.2. Physical Properties	63
5.3. Deterioration of Ceramic Materials	65
5.4. Carbons	67
Problems	70
Further Reading	72

Chapter 6 • POLYMERIC MATERIALS ... 73

6.1. Polymerization	74
6.1.1. Condensation Polymerization	74
6.1.2. Addition or Free Radical Polymerization	75
6.1.3. Solid State of Polymers	79
6.2. Effect of Structural Modification on Properties	82
6.2.1. Effect of Molecular Weight and Composition	82
6.2.2. Effect of Side-Chain Substitution, Cross-Linking, and Branching	83
6.3. Properties of Polymers	84
6.3.1. Mechanical Properties	85
6.3.2. Thermal Properties	87
6.4. Deterioration of Polymers	88
6.4.1. Chemical Effects	88
6.4.2. Thermal Effects during Sterilization	88
6.4.3. Mechanochemical Effect	89
6.4.4. Deterioration of Polymers *in Vivo*	89
Problems	92
Further Reading	96

Chapter 7 • STRUCTURE–PROPERTY RELATIONSHIPS OF BIOLOGICAL MATERIALS ... 97

7.1. Structure of Proteins and Polysaccharides	98
7.1.1. Proteins	98
7.1.2. Polysaccharides	102
7.2. Structure–Property Relationship of Tissues	104
7.2.1. Collagen-Rich and Mineralized Tissues	104
7.2.2. Elastic Tissues	114

CONTENTS

Problems 121
Further Reading 129

Chapter 8 • TISSUE RESPONSE TO IMPLANTS 131

8.1. Wound Healing Process 131
    8.1.1. Inflammation 131
    8.1.2. Cellular Response to Repair 133
8.2. Body Response to Implants 138
    8.2.1. Cellular Response to Implants 138
    8.2.2. Systemic Effects by Implants 142
Problems 144
Further Reading 146

Chapter 9 • SOFT TISSUE REPLACEMENT I: SUTURES, SKIN, AND MAXILLOFACIAL IMPLANTS 147

9.1. Sutures, Surgical Tapes, and Adhesives 148
    9.1.1. Sutures 148
    9.1.2. Surgical Tapes 149
    9.1.3. Tissue Adhesives 150
9.2. Percutaneous and Skin Implants 152
    9.2.1. Percutaneous Devices 152
    9.2.2. Artificial Skin 157
9.3. Maxillofacial and Other Soft Tissue Augmentation 158
    9.3.1. Maxillofacial Implant 159
    9.3.2. Other Soft Tissue Implants 159
Problems 160
Further Reading 161

Chapter 10 • SOFT TISSUE REPLACEMENT II: BLOOD INTERFACING IMPLANTS 163

10.1. Blood Compatibility 164
    10.1.1. Factors Affecting Blood Compatibility 165
    10.1.2. Nonthrombogenic Surfaces 165
10.2. Implants for Blood Interface 168
    10.2.1. Vascular Implants 168
    10.2.2. Heart Valve Implants 170
    10.2.3. Heart Assist Devices 172
    10.2.4. Artificial Organs 175
Problems 182
Further Reading 184

Chapter 11 • HARD TISSUE REPLACEMENT I: LONG BONE REPAIR ... 187

11.1. Internal Fracture Fixation Devices ... 188
    11.1.1. Wires and Screws ... 188
    11.1.2. Fracture Plates ... 190
    11.1.3. Intramedullary Devices ... 194
    11.1.4. Nail and Plate Devices for Femoral Osteotomy ... 195
    11.1.5. Spinal Fixation Devices ... 198
11.2. Materials Used for Internal Fracture Fixation Devices ... 200
    11.2.1. Stainless Steels ... 201
    11.2.2. Cobalt–Chromium Alloys ... 202
    11.2.3. Other Metals ... 204
Problems ... 205
Further Reading ... 207

Chapter 12 • HARD TISSUE REPLACEMENT II: JOINTS AND TEETH ... 209

12.1. Structure and Function of Joints ... 210
12.2. Various Joint Replacements ... 211
    12.2.1. Hip Joint Replacement ... 212
    12.2.2. Other Joint Replacements ... 220
    12.2.3. Materials Used for Joint Replacements ... 220
12.3. Dental Implants ... 226
    12.3.1. Endosseous Implants ... 226
    12.3.2. Other Dental Implants ... 228
    12.3.3. Materials Used for Dental Implants ... 228
Problems ... 230
Further Reading ... 231

APPENDIX: SI UNITS ... 233

NAME INDEX ... 235

SUBJECT INDEX ... 239

CHAPTER 1

INTRODUCTION

The word *biomaterials* can be interpreted in two ways—first, as such biological materials as tissues and wood; and second, as implant materials that replace the function of the biological materials. According to its legal definition (Clemson Advisory Board for Biomaterials "Definition of the word 'Biomaterials,'" the Sixth Annual International Biomaterial Symposium, April 20–24, 1974), "a biomaterial is a systemically, pharmacologically inert substance designed for implantation within or incorporation with living systems." This definition clearly emphasizes biomaterial as an implant material, although the conventional usage of the prefix "bio-" (in biochemistry, biophysics, and bioassay, for example) is violated. To avoid confusion I will use the term "biomaterials" to mean implants replacing biological materials. In this definition, implantable biomaterial includes anything that is intermittently or continuously exposed to body fluids even though it may be actually located outside the body proper. It includes most dental materials, although traditionally they have been treated as separate entities. Such devices as external artificial limbs, hearing aids, and external facial "prostheses" are not implants.

Because the ultimate goal of using biomaterials is to restore function of the natural tissues and organs in the body, it is very important to understand the relationships among property, function, and structure for biological materials. Thus one can envision three areas in the study of biomaterials—biological materials, implant materials, and interaction between the two in the body. These relationships are difficult to master unless one has fundamental knowledge of the entire system.

Another important area of study is that of the mechanics and dynamics of tissues and the interactions among them. Generally, this kind of study, called biomechanics, is incorporated in the design and insertion of an implant (Fig. 1-1).

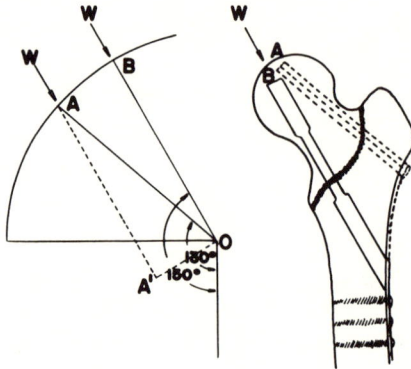

Figure 1-1. A biomechanical analysis of femoral neck fracture fixation. If the implant is positioned at 130° rather than 150° there will be a force component that will generate a bending moment at the nail-plate junction. Because the 150° implant is harder to insert it is not preferred by surgeons (modified from W. K. Massie, *J. Bone Joint Surg., 46A*, 684, 1964).

The performance of an implant after insertion can be considered in terms of its reliability. For example, four major factors contribute to the failure of hip joint replacements: fracture, wear, loosening of the implants, and infection. If we assume the probability of failure of a given system is f then the reliability r can be expressed as

$$r = 1 - f \qquad (1\text{-}1)$$

The total reliability r_t can be expressed in terms of the reliabilities of absence of each contributing factor for failure,

$$r_t = r_1 \cdot r_2 \cdots r_n \qquad (1\text{-}2)$$

where $r_1 = 1 - f_1$, $r_2 = 1 - f_2$, and so on.

Equation (1-2) implies that even though the implant is perfect (i.e., $r = 1$), if an infection occurs every time it is implanted then its total reliability is zero.

Example 1-1

Calculate the reliability of a total hip replacement operation if the probabilities of infection, loosening, wear, and fracture failure are 2, 3, 1, and 5%, respectively.

INTRODUCTION

Answer

$$r_t = (1 - 0.02)(1 - 0.03)(1 - 0.01)(1 - 0.05)$$
$$= 0.89 \text{ (1 in 10 procedures will } not \text{ be satisfactory)}$$

In this case I've ignored the time element because the loosening, wear, fracture, and infection are time-dependent factors, as shown in Figure 1-2.

The study of the structure–property relationship of biological materials is as important as that of biomaterials; however, traditionally this subject has not been treated fully in biologically oriented disciplines because most of them are interested in biological function rather than viable or nonviable "materials" as such. In many cases biological materials can be studied ignoring the fact that they are made of and made from living cells. In other cases the function of the tissues or organs is so vital that it is useless to replace them with biomaterials—for example, spinal cord or the brain. In this case I will treat the subject as if it does not have living cells unless this is an important factor contributing to the properties or function of the materials in the body.

The success of a biomaterial or implant is highly dependent on three major factors: the properties and biocompatibility of the implant, the condition of the recipient, and the competency of the surgeon who implants and monitors progress of the implant. It is easy to understand the requirements of the implant by imagining the needs for a bone plate to stabilize a fractured femur after an accident:

1. Acceptance of the plate to the tissue surface—that is, biocompatibility [this is a broad term and sometimes includes (2) and (3) of this list]

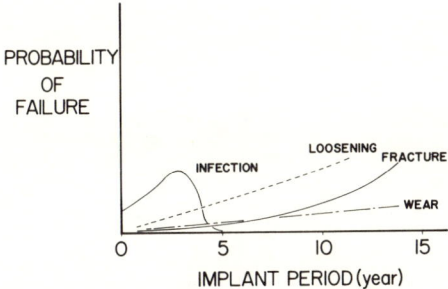

Figure 1-2. A schematic illustration of probability of failure versus implant period for hip joint replacements (redrawn from J. H. Dumbleton, *J. Med. Eng. Tech.*, 1(6), 341, 1977).

Table 1-1. Materials for Implantation

Materials	Advantages	Disadvantages	Examples
Polymers Silastic® rubber Teflon® Dacron® Nylon	Resilience, easy to fabricate, low density	Low mechanical strength, time-dependent degradation	Sutures, arteries, veins; maxillofacial—nose, ear, maxilla, mandible, teeth; cement, artificial tendons
Metals 316, 316L S.S. Vitallium® Titanium alloys	High-impact tensile strength, high resistance to wear, ductile adsorption of high strain energy	Low biocompatibility, corrosion in physiological environment, mismatch of mechanical properties with soft connective tissues, high density	Orthopedic fixation—screws, pins, plates, wires; intermedullary rods, staples, nails; dental implants
Ceramics Aluminum oxides Calcium aluminates Titanium oxides Carbons	Good biocompatibility, corrosion resistance, inert, high compression resistance	Low tensile impact strength, difficult to fabricate, low mechanical reliability, lack of resilience, high density	Hip prosthesis, ceramic teeth, transcutaneous device
Composites Ceramic-coated metal Carbon-coated material	Good biocompatibility, inert, corrosion resistance, high tensile strength	Inconsistent material fabrication	Artificial heart valve (pyrolytic carbon on graphite), knee joint implants (carbon-fiber-reinforced high-density polyethylene)

INTRODUCTION

2. Nontoxic and noncarcinogenic
3. Chemically inert and stable (non-time-dependent degradation)
4. Adequate mechanical strength
5. Adequate fatigue life
6. Sound engineering design
7. Proper weight and density
8. Relatively inexpensive, reproducible, and easy to fabricate and process on a large scale

Most of these requirements are pertinent to the material properties that are the subjects of this book. Table 1-1 illustrates some of the advantages, disadvantages, and applications for four groups of synthetic (man-made) materials used for implantation.

Reconstituted (natural) materials such as collagen have been tried experimentally with a limited success for such applications as arterial wall, heart valve, or skin.

Another alternative to the artificial implant is a natural transplantation, such as the kidney and heart, but this effort has been stymied because of social, ethical, and immunological problems. However, in case of kidney failure, the patient's hope lies in transplantation because an artificial kidney has many disadvantages, such as high cost, immobility, and constant care and maintenance of the dialyzer.

The surgical uses of implant materials have been classified in the following list for quick reference:

Surgical uses of biomaterials

1. Permanent implant
   a. Muscular skeleton system—elbow, knee, hip, shoulder, finger and toe bone segment, permanently attached artificial limb
   b. Cardiovascular system—heart (valve, wall, pacemaker, entire heart), arteries, veins
   c. Respiratory system—larynx, trachea and bronchus, chest wall, diaphragm, lungs, thoracic plombage
   d. Digestive system—esophagus, bile ducts, liver
   e. Genitourinary system—kidney, ureter, urethra, bladder
   f. Nervous system—dura, hydrocephalus shunt
   g. Special senses—corneal and lens prosthesis, ear, carotid pacemaker
   h. Other soft tissues—hernia, tendons, visceral adhesion
   i. Cosmetic implants—maxillofacial (nose, ear, maxilla, mandible, teeth), breast, eye, testes, penis, etc.

2. Transient implant
    a. Extracorporeal assumption of organs' functions—heart, lungs, kidney, liver
    b. Decompressive (drainage of hollow viscera spacings)—gastrointestinal (biliary), genitourinary, thoracic, peritoneal lavage, cardiac catheterization
    c. External dressings and partial implants—temporary artificial skin, immersion fluids
    d. Aids to diagnosis—catheters, probes
    e. Orthopedic fixation devices—general (screws, hip pins, traction); bone plates (long bone, spinal, oseotomy); intertrochanteric (hip nail, nail-plate combination, threaded or unthreaded wires and pins); intramedullary (rods and pins); staples
    f. Sutures and surgical adhesives

FURTHER READING

J. H. U. Brown, J. E. Jacobs, and L. Stark, *Biomedical Engineering,* chapter 11, F. A. Davis Co., Philadelphia, 1971.

L. Stark and G. Agarwal (eds.), *Biomaterials,* Plenum Press, New York, 1969.

S. A. Wesolowski, A. Martinez, and J. D. McMahon, *Use of Artificial Materials in Surgery,* Year Book Medical Publishers, Chicago, 1966.

D. F. Williams and R. Roaf, *Implants in Surgery,* chapter 1, W. B. Saunders, London, 1973.

CHAPTER 2

CHARACTERIZATION OF MATERIALS

2.1. MECHANICAL PROPERTIES

2.1.1. Stress–Strain Behavior

If a material is undergoing a mechanical deformation, the force per unit area is defined as a *stress*, which is usually expressed in newtons per square meter (pascal, Pa):

$$\text{stress}\ (\sigma) = \frac{\text{force}}{\text{cross-sectional area}} \quad \left[\frac{\text{N}}{\text{m}^2}\right] \qquad (2\text{-}1)$$

Three types of stress will be present singly or in combination whenever a load is placed on a material: tension, compression, and shear stresses. Tensile stresses are generated in response to loads which pull an object apart (Fig. 2-1a), while compressive stresses resist crushing loads (Fig. 2-1b). Shear stresses resist loads that deform or separate by sliding layers of molecules past each other on one or more planes (Fig. 2-1c). The shear stresses can also be found in uniaxial tension or compression since the applied stress produces maximum stress on planes at 45° to the direction of loading (Fig. 2-1d).

The deformation of an object in response to an applied load is called strain:

$$\text{strain}\ (\epsilon) = \frac{\text{elongated length} - \text{original length}}{\text{original length}} \quad \left[\frac{\text{m}}{\text{m}}\right] \qquad (2\text{-}2)$$

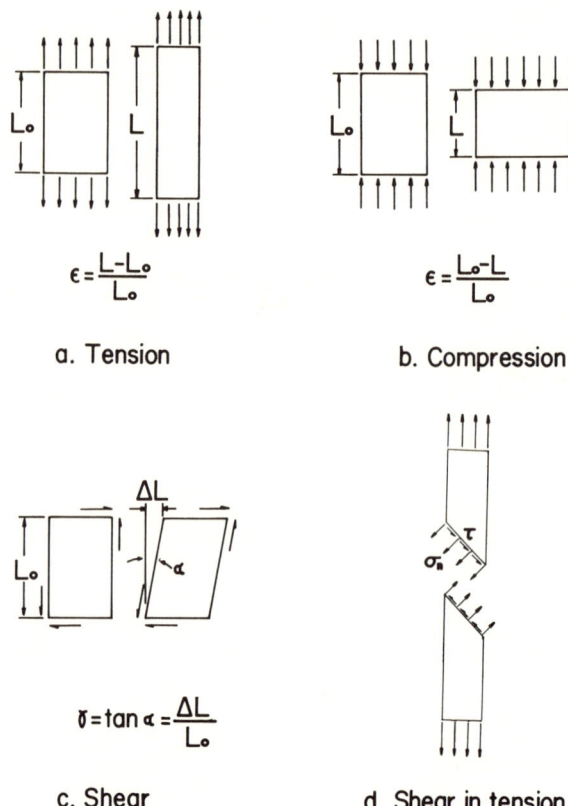

Figure 2-1. Three different modes of deformation. The shear stresses can be produced by tension or compression as in d.

It is also possible to denote strain as the stretch ratio—that is, elongated length/original length. The deformation associated with different types of stresses are called tensile, compressive, and shear strain (cf. Fig. 2-1).

If the stress–strain behavior is plotted on a graph, a curve that represents a continuous response of the material toward the imposed force is seen (Fig. 2-2). The stress–strain curve of a solid is divided by the yield point (YS) into elastic and plastic regions. In the elastic region, the strain increases in direct proportion to the applied stress (Hooke's law):

$$\sigma = E\epsilon \qquad (2\text{-}3)$$

The slope E is called Young's modulus or the modulus of elasticity. It is the value of the stress–strain ratio. The stiffer a material is, the higher the

CHARACTERIZATION OF MATERIALS

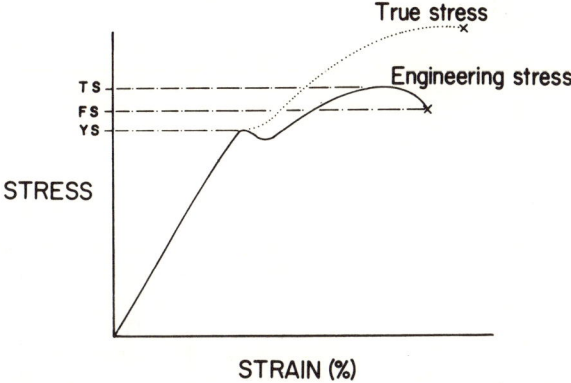

Figure 2-2. Stress–strain behavior of an idealized material.

value of E and the more difficult it is to deform. The unit for the modulus is the same as that of stress because strain is dimensionless.

In the plastic region, strain changes are no longer proportional to the applied stress. Further, when the applied stress is removed, the material will not return to its original shape but will be permanently deformed, which is called plastic deformation.

Figure 2-3 is a schematic illustration of what will happen when a material is strained elastically. The individual atoms are distorted and stretched, while part of the strain is accounted for by a limited movement of atoms past one another. When the load is released the atoms will go back to their original configuration. When a material is deformed plastically (Fig. 2-3), the atoms are moved past each other in such a way that they will have new neighbors; when the load is released they can no longer go back to their original positions.

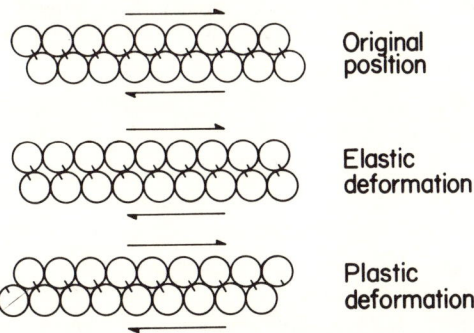

Figure 2-3. A two-dimensional atomic model showing elastic and plastic deformation.

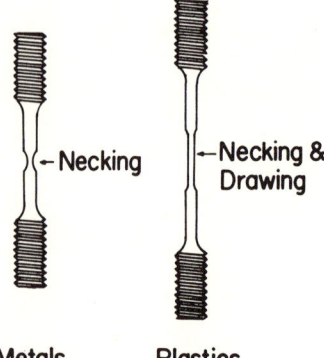

Figure 2-4. Deformation characteristics of metals and plastics under stress. Note that metals rupture without further elongation after necking occurs, e.g., in plastics, the necked region undergoes further deformation called drawing.

The peak stress in Figure 2-2 is often followed by an apparent decrease until a point is reached where the material ruptures. The peak stress is known as the *tensile* or *ultimate tensile strength* (TS); the stress where failure occurs is called the *failure* or *rupture strength* (FS).

In many materials such as stainless steels, definite yield points occur. This point is characterized by temporarily increasing strain without further increase in stress. Sometimes, when it is difficult to decipher the yield point for a given stress–strain curve, an offset yield point (usually 0.2%) is used in lieu of the original yield point.

Thus far examples of engineering stress–strain curves have been examined. These differ from the true stress–strain curves in the following particular, the cross-sectional area over which the load is acting is assumed to be constant from the initial loading until final rupture. That this assumption is not correct accounts for the peak seen at the ultimate tensile point. For example, as a specimen is loaded in tension, necking sometimes occurs (Fig. 2-4), which changes the area over which the load is acting. If adjustments are made for the changes in cross-sectional area, then a dotted curve like that in Figure 2-2 is obtained.

Example 2-1

The following curve was obtained by stretching an electrical wire (Cu) with a diameter of 1.30 mm.

a. Calculate modulus of elasticity and tensile strength.
b. Calculate the yield strength.
c. Calculate the true fracture strength if the final diameter of the wire just before break was 1.20 mm.

CHARACTERIZATION OF MATERIALS

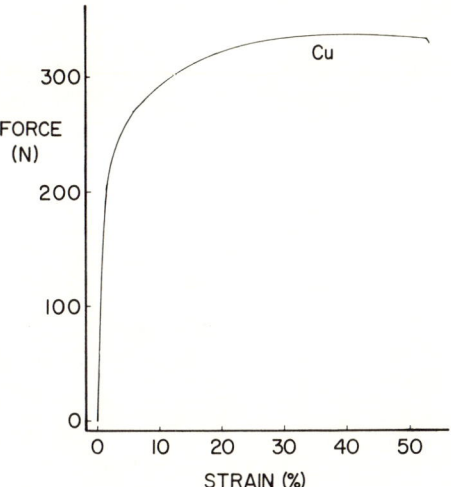

Force versus strain for a copper wire.

Answer

a. Tensile strength

$$\sigma = \frac{342 \text{ N}}{\pi (0.65)^2 \times 10^{-6} \text{ m}^2} \doteq \underline{258 \text{ MPa}}$$

Modulus of elasticity

$$E = \frac{\sigma}{\epsilon} = \frac{258 \text{ MPa}}{1.27\%} = \underline{20.3 \text{ GPa}}$$

(This value is about ¼ of the value due to "machine" deformation reported in the literature.)

b. $$\text{Yield strength} = \frac{105 \text{ N}}{\pi (0.65)^2 \times 10^{-6} \text{ m}^2} = \underline{79.1 \text{ MPa}}$$

c. $$\text{True tensile strength} = \frac{342 \text{ N}}{\pi (0.60)^2 \times 10^{-6} \text{ m}^2} = \underline{302 \text{ MPa}}$$

This example illustrates the difficulties involved when attempting to determine exact values working from the somewhat arbitrary values assigned by researchers.

2.1.2. Mechanical Failure

Mechanical failure occurs usually by fracture. The fracture of a material can be characterized by the amount of energy required to produce the failure. The quantity is called toughness and can be expressed in terms of stress and strain:

$$\text{Toughness (energy)} = \int_{\epsilon_o}^{\epsilon_f} \sigma \, d\epsilon = \int_{l_o}^{l_f} \sigma \, dl/l \qquad (2\text{-}4)$$

Expressed another way, toughness is the summation of stress times the distance over which it acts (strain) taken in small increments. The area under the stress–strain curve provides a simple method of estimating toughness (Fig. 2-5).

A material that can withstand high stresses and will undergo considerable plastic deformation (ductile-tough material) is tougher than one which resists high stresses but has no capacity for deformation (hard brittle material) or one which has a high capacity for deformation but can only withstand relatively low stresses (ductile-soft or plastic material).

The two major characteristics of brittle fracture are that its fracture stress is far below the theoretical strength and that it is difficult to predict. The latter fact is the major reason why ceramic and glassy materials are not used extensively for implantation despite their excellent compatibility with tissue. The reason for the lower fracture stress of brittle materials is discussed in Chapter 5.

Like toughness, impact strength is the amount of energy that can be absorbed by a material but the force is applied by impact. It can be measured by subjecting the specimen of known dimensions to a swinging pendulum. The amplitude change of the swing of the pendulum is the measure of the energy absorbed by the specimen. From this the impact

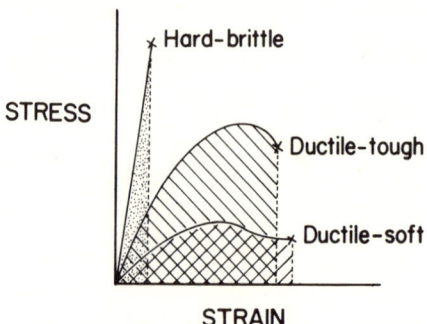

Figure 2-5. Stress–strain curves of different types of materials. The areas underneath the curves are the measure of toughness.

CHARACTERIZATION OF MATERIALS

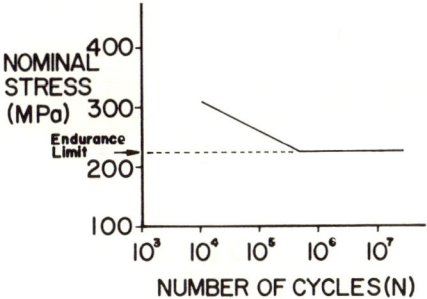

Figure 2-6. Typical curve of a fatigue testing result plotted in nominal stress versus fatigue cycles.

strength or energy can be calculated. Usually the impact testing requires a large number of samples because there is a large variation in results.

When a material is subjected to a constant or a repeated load below fracture stress it can fail after some time. This is called static or dynamic (cyclic) fatigue and is usually plotted as stress versus time or log cycles (N), as shown in Figure 2-6. The time or the number of cycles before failure depends on the magnitude and types of load, the test environment, and temperature.

Cyclic fatigue is characteristic of ductile or plastic materials, although the final fracture is rather rapid. The reason for the cyclic fatigue is the inhomogeneity and anisotropy of materials. Imperfections, particularly those on the surface caused by machining and handling, can initiate cracks, and the growth of cracks leading to catastrophic failure can occur inside the material under cyclic loading. A material may undergo cyclic loading indefinitely below a certain stress level called the endurance limit.

Fatigue tests in a simulated body environment will give a better evaluation of the material because the materials placed in the body undergo loading and unloading cycles. However, it is almost impossible to simulate the complicated loading and unloading conditions an implant undergoes *in vivo*. Nevertheless, the fatigue test is useful for comparing performances of various implants under some testing conditions.

2.1.3. Viscoelasticity

Although the simple equation (2-3) can describe the elastic behavior of many materials at low strain as shown in Figure 2-7, it cannot be used to characterize the polymers and tissues that are the major concern of this book. The fluidlike behavior of a material (such as water and oil) can be described in terms of stress and strain as in the elastic solids, but the

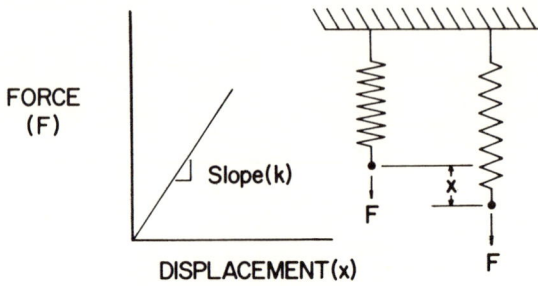

Figure 2-7. Force versus displacement of a spring.

proportionality constant (viscosity) is derived from the following relationship:

$$\sigma = \eta \, d\epsilon/dt \quad \text{(stress = viscosity} \times \text{strain rate)} \tag{2-5}$$

The stress and strain are shear rather than tensile or compressive, although the same symbols are used to avoid complications.

A mechanical analog (dashpot) can be used to simulate the viscous behavior of equation (2-5) as shown in Figure 2-8. A similar construction, an automobile shock absorber cylinder, contains oil as the damping fluid. Equation (2-5) shows that stress is time-dependent; i.e., if the deformation is accomplished in a very short time ($dt \to 0$) then the stress becomes infinite. On the other hand, if the deformation is achieved slowly ($dt \to \infty$), the stress approaches zero regardless of the viscosity value.

In principle, the simple equations (2-3) and (2-5) can describe the viscoelastic behavior of a material when combined as if the material is made of springs and dashpots. The stress–strain behavior of the spring and dashpot can be represented as shown in Figure 2-9. If the spring and

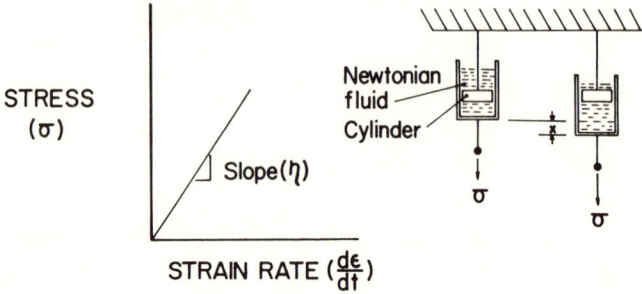

Figure 2-8. Stress versus strain rate of a dashpot.

CHARACTERIZATION OF MATERIALS   15

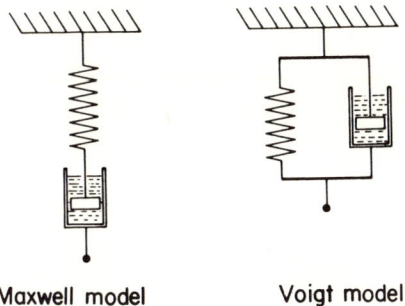

Figure 2-9. Two-element viscoelastic models.

dashpot are arranged in series and parallel they are called Maxwell and Voigt (or Kelvin) models, respectively. Remember that equation (2-3) does not involve time, implying that the spring acts instantaneously when stressed. Hence, if the Maxwell model is stressed suddenly the spring reacts instantaneously, while the dashpot cannot react since its piston cannot move because of the infinite stress required by the surrounding fluid. However, if we hold the Maxwell model after instantaneous deformation, the dashpot will react to the retraction of the spring and this will take time (dt = finite). This description can be expressed concisely by a simple mathematical formulation. In general, the response to stress by the Maxwell model will result on the strain as a cumulative; that is, total strain is a combination of the strain of spring and dashpot

$$\epsilon_{total} = \epsilon_{spring} + \epsilon_{dashpot} \tag{2-6}$$

Differentiating both sides

$$\frac{d\epsilon_{total}}{dt} = \frac{d\epsilon_{spring}}{dt} + \frac{d\epsilon_{dashpot}}{dt} \tag{2-7}$$

and rewriting equation (2-3)

$$\frac{d\sigma_{spring}}{dt} = E \frac{d\epsilon_{spring}}{dt} \tag{2-8}$$

and substituting equations (2-8) and (2-5) into (2-7), one gets

$$\frac{d\epsilon_{total}}{dt} = \frac{1}{E} \frac{d\sigma_{spring}}{dt} + \frac{\sigma_{dashpot}}{\eta} \tag{2-9}$$

Also, one can see that the total stress is the same for the spring and the dashpot because each member has to oppose the same applied load internally (or else it breaks!). Thus equation (2-9) becomes

$$\frac{d\epsilon}{dt} = \frac{1}{E}\frac{d\sigma}{dt} + \frac{\sigma}{\eta} \qquad (2\text{-}10)$$

Equation (2-10) can be applied easily for a simple mechanical test such as "stress relaxation" in which the specimen is strained instantaneously and the relaxation of the load is monitored while the specimen is held at a constant length. Thus the strain rate becomes zero ($d\epsilon/dt = 0$) and equation (2-10) can be written

$$\frac{1}{E}\frac{d\sigma}{dt} + \frac{\sigma}{\eta} = 0 \qquad (2\text{-}11)$$

Therefore,

$$d\sigma/dt = -E\sigma/\eta \qquad (2\text{-}12)$$

$$d\sigma/\sigma = -(E/\eta)\, dt \qquad (2\text{-}13)$$

and

$$\ln \sigma = -(E/\eta)t + \text{constant} \qquad (2\text{-}14)$$

Since at $t = 0$, $\sigma = \sigma_o$, thus $\ln \sigma_o = \text{constant}$, and

$$\frac{\sigma}{\sigma_o} = \exp\,[-(E/\eta)t] \qquad (2\text{-}15)$$

The constant η/E can be substituted with another constant τ called relaxation time and equation (2-15) will become

$$\sigma = \sigma_o \exp\,(-t/\tau) = \sigma_o/\exp\,(t/\tau) \qquad (2\text{-}16)$$

Examining equation (2-16) one can see that if the relaxation time is short then the stress (σ) at a given time becomes small. On the other hand, if the relaxation time is long then the stress (σ) is the same as the original stress (σ_o).

CHARACTERIZATION OF MATERIALS

Example 2-2

A stress of 1 MPa is required to stretch a 2-cm skin strip to 2.5 cm. After 1 h in the same stretched position, the strip exerted a stress of 0.5 MPa. Assume the property of skin does not vary appreciably during the experiment.

a. What is the relaxation time?
b. What stress would be exerted by the skin strip in the same stretched position after 5 h?

Answer

a. From equation (2-16)

$$\sigma = \sigma_o \exp(-t/\tau)$$

$$0.5/1 = \exp(-1/\tau)$$

therefore

$$\tau = 1.44 \text{ h}$$

b. $\sigma = 1 \exp(-5/1.44) = \underline{0.031 \text{ MPa}}$

In comparison, a window glass has a large relaxation time (no stress relaxation), while water and oil have short relaxation times. Thus when stressed their shape changes immediately to relieve the applied stress (instantaneous stress relaxation).

Similar analysis can be made with the Voigt model. In this case the strain of spring and dashpot represent the total strain, i.e.,

$$\epsilon_{total} = \epsilon_{spring} = \epsilon_{dashpot} = \epsilon \qquad (2\text{-}17)$$

The total stress is a cumulative of spring and dashpot

$$\sigma_{total} = \sigma_{spring} + \sigma_{dashpot} = \sigma \qquad (2\text{-}18)$$

Substituting equations (2-3) and (2-5) into (2-18)

$$\sigma = E\epsilon + \eta \, d\epsilon/dt \qquad (2\text{-}19)$$

If a stress is applied and the stress is removed after a certain time, then

$$0 = E\epsilon + \eta \, d\epsilon/dt \qquad (2\text{-}20)$$

which is similar to equation (2-11) and can be solved similarly; hence

$$\epsilon_{\text{recovery}} = \epsilon_o \exp\left[-(E/\eta)t\right] \quad (2\text{-}21)$$

where ϵ_o is the strain at the time of stress removal. The constant η/E is termed retardation time (λ) for this creep recovery process. Because the strain is being recovered from the original strain (ϵ_o), equation (2-21) can be rewritten:

$$\epsilon = \epsilon_o - \epsilon_o \exp(-t/\lambda) = \epsilon_o[1 - \exp(-t/\lambda)] \quad (2\text{-}22)$$

Example 2-3

A piece of polymer (polypropylene) is stretched 10% of its length. When the tension is released, it recovers 50% of its strain after one day at room temperature.

a. What is the retardation time?
b. What is the amount of strain recovered after 10 days at room temperature?

Answer

a. From equation (2-22)

$$\epsilon = \epsilon_o[1 - \exp(-t/\lambda)]$$

$$\epsilon/\epsilon_o = 0.5 = 1 - \exp(-1/\lambda)$$

Therefore

$$\lambda = 1.443 \text{ days}$$

b. $\epsilon = 0.1[1 - \exp(-10/1.443)]$
   $= 0.0999$
   $= \underline{9.99\%}$ (this is 99.9% recovery of strain since the original strain is 10%)

2.1.4. Pure Bending Stress

When a simple beam supported by a pin and a roller is subjected to bending, as shown in Figure 2-10, then the longitudinal element in the convex side of the beam will be elongated (tension) while the concave side element will be shortened (compression). The element that does not elongate or shorten is called the neutral axis (NA) of the beam. An example of

CHARACTERIZATION OF MATERIALS

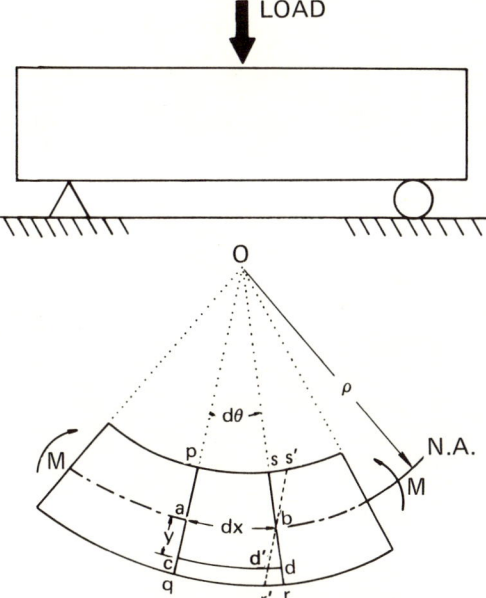

Figure 2-10. Bending of a beam.

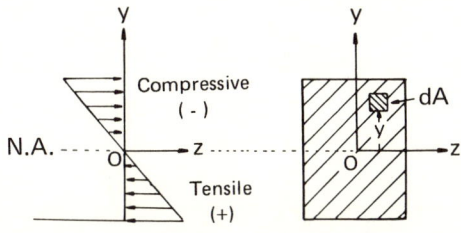

Figure 2-11. Cross-section of a bent beam (cf. Fig. 2-10).

the internal stress distribution is shown for a rectangular cross section in Figure 2-11.

The two-dimensional element $pqr's'$ from the beam will be changed to $pqrs$ after bending (Fig. 2-10b); let ρ be the curvature of the neutral axis, then the small element $abcd$ can be constructed. The $r's'$ line is parallel to the pq line before bending and dd' distance can be expressed as

$$dd' = y\, d\theta \tag{2-23}$$

Since $ab = cd' = dx = \rho\, d\theta$, the strain is

$$\epsilon_x = dd'/cd' = y\, d\theta/dx = y/\rho \tag{2-24}$$

The strain can be negative if we consider the inner fiber of the bent beam. By using the simple relationship of stress–strain if the material behaves as a spring (Hooke's law), then

$$\sigma_x = \epsilon_x E = Ey/\rho \tag{2-25}$$

Let dA denote a small area of the cross section at the distance y from the neutral axis, as shown in Figure 2-11. The force element on this area becomes $\sigma_x\, dA$ and from equation (2-25)

$$\sigma_x\, dA = Ey dA/\rho \tag{2-26}$$

The moment of the force element $\sigma_x\, dA$ about the neutral axis is given as

$$dM = y\sigma_x\, dA \tag{2-27}$$

The total moment can be calculated:

$$M = \int_A y\sigma_x\, dA = (E/\rho) \int_A y^2\, dA \tag{2-28}$$

The integral in this equation is called the moment of inertia of the cross-sectional area with respect to that axis and expressed as I. Combining equations (2-25) and (2-28),

$$M/I = \sigma_x/y \tag{2-29}$$

For the maximum stress y is replaced with c, which is defined as the distance from the neutral axis to the remotest element. Thus equation (2-29) can be rewritten as

$$\max \sigma_x = M/(I/c) = M/Z \tag{2-30}$$

The section moduli ($Z = I/c$) of some cross sections are given in Figure 2-12. The reader should be able to calculate the advantages of having a hollow cross section based on the section modulus calculations. Another relation can be derived easily:

$$M = EI/\rho \tag{2-31}$$

CHARACTERIZATION OF MATERIALS

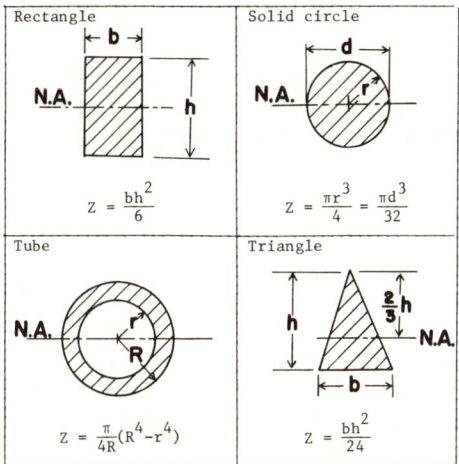

Figure 2-12. Section modulus of various cross sections.

where E is the modulus of elasticity and ρ is the radius of bending. The relation predicts that for the maximum bending moment one needs to have a material with a high modulus of elasticity (stiff).

Example 2-4

During hip replacement surgery sometimes the greater trochanter is removed for easier insertion of the femoral implant and later a Kirschner wire is used to reattach the bone. If the wire has a 1-mm diameter and the minimum bending curvature of the wire has a 5-cm radius, calculate the maximum bending stress that will be produced in the wire made of a wrought Co–Cr alloy (fully annealed).

Answer

From equation (2-25) and Table 11-6,

$$\sigma_x = Ey/\rho$$
$$= \frac{(230 \times 10^9 \text{ Pa})(0.5 \times 10^{-3} \text{ m})}{5.0 \times 10^{-2} \text{ m}}$$
$$= 2.30 \times 10^9 \text{ Pa (or 2.30 GPa)}$$

This value is well over the tensile strength of the material, thus the outer surface of the wire will break! This is the reason these wires are made up of many fine wires, i.e., multistrands.

2.2. THERMAL PROPERTIES

The most familiar thermal properties are the melting and freezing (solidification) temperatures; that is, the phase transformation occurs at some specific temperatures. These temperatures depend on the bond energy—for example, the higher the bonding strength the higher the melting temperature. If the material is made of many elements or different molecules, then it may have a range of melting or solidification temperatures.

The thermal energy spent on converting one gram of a material from solid to liquid is called heat of fusion. The unit is joules per gram, where one joule is equivalent to one newton meter. The heat of fusion is closely related to the melting temperature (T_m)—i.e., the higher the T_m, the higher the heat of fusion, although there are many exceptions (Table 2-1).

The thermal energy spent on changing 1°C of a material for a unit mass is called specific heat. Traditionally water is usually chosen as a standard substance and 1 calorie is the heat required to raise 1 g of water from 15°C to 16°C, but now the standard unit of heat is the joule, and the specific heat is J/g (1 calorie is equivalent to 4.187 J).

Table 2-1. Thermal Properties of Materials

Substances	Melting temp. (°C)	Specific heat (J/g)	Heat of fusion (J/g)	Thermal conductivity (W/mK)	Linear coefficient of expansion ($\times 10^{-6}$/°C)
Mercury	−38.87	0.138	12.7	68	60.6
Gold	1063	0.13	67	297	14.4
Silver	960.5	0.2345	108.9	421	19.2
Copper	1083	0.385	205.2	384	16.8
Platinum	1773	0.134	113	70	—
Enamel	—	0.75	—	0.82	11.4
Dentine	—	1.17	—	0.59	8.3
Acrylic resin	70 [a]	1.465	—	0.2	81.0
Water	0	4.187	334.9 (ice)	—	—
Paraffin	52	2.889	146.5	—	—
Beeswax	62	—	175.8	0.4	350
Alcohol	−117	2.29	104.7	—	—
Glycerin	18	2.428	75.4	—	—
Amalgam	480	—	—	23	22.1–28
Porcelain	—	1.09	—	1	4.1

[a] Softening temperature.

CHARACTERIZATION OF MATERIALS

The change in length for a unit length by thermal energy is called linear coefficient of expansion (α) which can be expressed

$$\alpha = \frac{\Delta l}{\text{original } l \cdot \Delta T} \left[\frac{m}{mK}\right] \tag{2-32}$$

The value may vary with the crystal direction of the material and temperature range in which it was measured. If the material is homogeneous then the volumetric thermal expansion coefficient (ν) can be approximated:

$$\nu \simeq 3\alpha \tag{2-33}$$

Another important thermal property is thermal conductivity, which is defined as the amount of heat passed for a given time, thickness, and area of the material. The unit is watts/mK where 1 W is equivalent to 1 J/s. Generally, thermal conductivity of metals is much higher than ceramics and polymers because the free electrons in metals act as energy conductors.

Example 2-5

To fill a cavity, a cylindrical hole with a 2-mm diameter is made in a molar tooth (there was no enamel left). The length of the hole is 4 mm and filled with amalgam and acrylic resin.

a. Calculate the volume changes for the fillings.
b. Calculate the force developed between the dentine and the fillers. Assume the temperature variation is 50°C. The moduli of elasticity of amalgam and resin are 20 GPa and 2.5 GPa, respectively.

Answer

a. Because the volume expansion coefficient ν can be defined as in equation (2-32)

$$\Delta V/V_o \Delta T = \nu = 3\alpha, \text{ therefore } \Delta V = V_o \cdot 3\alpha \cdot \Delta T$$

The net volume changes after filling will be

$$\Delta V_{\text{amalgam}} = V_o \times 3(\alpha_{\text{amalgam}} - \alpha_{\text{dentin}}) \Delta T$$
$$= \pi (1 \text{ mm})^2 \times 4 \text{ mm} \times 3(25 - 8.3) \times 10^{-6} \times 50$$
$$= \underline{0.03 \text{ mm}^3}$$
$$\Delta V_{\text{resin}} = \pi (1 \text{ mm})^2 \times 4 \text{ mm} \times 3(81 - 8.3) \times 10^{-6} \times 50$$
$$= \underline{0.14 \text{ mm}^3}$$

b. Because $F = \sigma A$, $A = \pi D h = \pi \times 2 \times 4 \times 10^{-6}$ m^2 = 25.13 × 10^{-6} m^2 and

$$\sigma = E \cdot \Delta \epsilon$$
$$\Delta \epsilon = \Delta T (\alpha_{\text{amalgam or resin}} - \alpha_{\text{dentin}})$$
$$F_{\text{amalgam}} = 25.13 \times 10^{-6} \times 20 \times 10^9 \times 16.7 \times 10^{-6} \times 50 = \underline{420 \text{ N}}$$
$$F_{\text{resin}} = 25.13 \times 10^{-6} \times 2.5 \times 10^9 \times 72.7 \times 10^{-6} \times 50 = \underline{228 \text{ N}}$$

Note that although the resin expands more than four times in volume compared to the amalgam, the force exerted on the tooth is half that of amalgam. The actual force exerted by the fillings will be much smaller than the calculated values because the fillings can expand freely toward the top of the holes. The force felt by using amalgam as a filler would be felt sooner than that due to the resin because the amalgam can conduct heat much faster than the resin.

2.3. SURFACE PROPERTIES AND ADHESION

Surface properties are important in any materials-related problems. The surface property is directly related to the bulk property because the surface is the discontinuous boundary between different phases. If ice is being melted then two surfaces are created between three phases, i.e., liquid (water), gas (air), and solid (ice).

Surface tension develops near the phase boundaries because the equilibrium-bonding arrangements are disrupted, leading to excess energy which will minimize the surface area. Other means of minimizing the surface energy include attracting foreign materials (adsorption) or bonding with the adsorbate (chemisorption).

The conventional units used to describe surfaces are dynes per cm or ergs per cm^2 for surface energy (or tension), but these units are exactly the same because 1 erg is 1 dyne cm. The SI unit is N/m as given in Table 2-2.

If a liquid is dropped on a solid surface then the liquid drop will

Table 2-2. Surface Tension of Materials

Substance	Temperature (°C)	Surface tension (N/m)
Lead	327	0.452
Mercury	20	0.465
Zinc	419	0.758
Copper	1131	1.103
Gold	1120	1.128

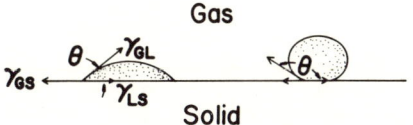

Figure 2-13. Wetting and nonwetting of a liquid on the flat surface of a solid. Note the contact angle (θ).

spread or make a spherical bubble, as shown in Figure 2-13. At equilibrium, the surface tension between the three phases in the solid plane should be zero because the liquid is free to move until force equilibrium is established.

$$\gamma_{GS} = \gamma_{LS} - \gamma_{GL} \cos \theta = 0$$
$$\gamma_{GS} = \gamma_{LS} + \gamma_{GL} \cos \theta \qquad (2\text{-}34)$$

where θ is called contact angle and the wetting characteristic can be generalized as

$$\begin{array}{ll} \theta = 0 & \text{complete wetting} \\ 0 < \theta < 90° & \text{partial wetting} \\ \theta > 90° & \text{nonwetting} \end{array} \qquad (2\text{-}35)$$

Note that equation (2-34) gives only ratios rather than absolute values of surface tension; some values of contact angle are given in Table 2-3.

Welding two surfaces together is called adhesion if the two surfaces are different and cohesion if they are the same. Hence all surfaces cemented with a cementing agent are demonstrating adhesion, and the cementing agent is an adhesive. For maximum welding strength the thickness of the adhesive layer must be optimal, as shown in Figure 2-14.

In dental and medical applications the adhesives should be considered a temporary remedy; because the tissues are living, they replace the old cells with new ones thus destroying the initial attachment. This prob-

Table 2-3. Contact Angle Values

Liquid	Substrate	Contact angle
Methylene iodine	Soda-lime glass	29°
CH_2I_2	Fused quartz	33°
Water	Paraffin	107°
Mercury	Soda-lime glass	140°

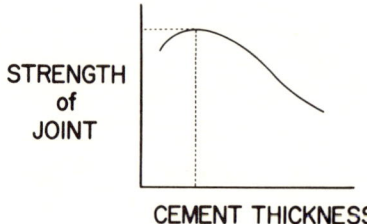

Figure 2-14. Variation of the strength of a joint versus thickness of cement between the adherend.

lem led to the development of porous implants that allow tissues to grow into the interstices (pores) making a viable interlocking system between implants and tissues.

PROBLEMS

2-1. Which of the following materials will fit closely the three stress–strain curves of Figure 2-5?

a. Ceramics and glasses
b. Plastics (polymer) such as polyethylene
c. Glassy polymers such as Plexiglass® (polymethylmethacrylate)
d. Soft tissues such as skin or blood vessel walls
e. Hard mineral tissues of bone and teeth
f. Steels
g. Rubber bands

Answers
Hard–brittle: a, c, e
Ductile–tough: b, f
Ductile–soft: d, g

2-2. Poisson's ratio (ν) is defined by the following expression for a cubic isotropic material:

$$\nu = -\frac{\epsilon_x}{\epsilon_z} = -\frac{\epsilon_y}{\epsilon_z}$$

where ϵ is strain and x, y, and z are directions. Determine Poisson's ratio of a cubic bone (wet) which has 1.000 cm³ of volume and is compressed hydrostatically to 0.999 cm³ with a 5 MPa pressure.

CHARACTERIZATION OF MATERIALS

Answer
From Figure 7-8 the strain is 0.36% at 5 MPa stress, thus $\epsilon_z = -0.0036$ and

Therefore,
$$0.999 = (1 - 0.0036)(1 + \epsilon_x)^2$$

$$\epsilon_x = 0.0013$$

and
$$\nu = \frac{-0.0013}{-0.0036}$$

$$= \underline{0.36}$$

2-3. An applied strain of 0.5 produces an immediate stress of 2 MPa in a piece of rubber, but after 10 days, the stress is only 1 MPa.

a. What is the relaxation time?
b. What is the stress after 100 days?

Answers
a. From equation (2-16)

$$1 = 2 \exp(-10/\tau)$$

$$\tau = \underline{14.4 \text{ days}}$$

b.
$$\sigma = 2 \exp(-100/14.4)$$

$$= \underline{0.002 \text{ MPa}}$$

2-4. Calculate the maximum diameter of the individual wire in Example 2-4 necessary to prevent breakage of the outer surface of the wire.

Answer
From equation (2-25) and Table 11-4

$$y = \frac{\sigma_x \rho}{E}$$

$$= \frac{(900 \times 10^6 \text{ Pa})(5 \text{ cm})}{230 \times 10^9 \text{ Pa}}$$

$$= \underline{0.02 \text{ cm}} \text{ (or 0.04 cm diameter)}$$

This indicates that if the outside surface of the wire is not to be broken, the wire diameter must be very small.

2-5. The surface properties change after a material is implanted inside the body. Explain how they will be changed and what methods should be used to understand the interaction between the tissue and the implant. Can you use the data obtained from *in vitro* surface experiments *in vivo?*

Answer
As soon as a material is in contact with the tissue it is covered with protein and lipid films masking the original surface properties.

2-6. What is the endurance limit for a solid polymer (polymethylmethacrylate) in Figure 6-4? If a tooth implant is made of this material with a 4-mm diameter in cylindrical shape, how long will it last? Assume that the average force of chewing is 30 N.

Answer

About 2 MPa

$$\sigma = \frac{30}{\pi(2 \times 10^{-3})^2} = 2.39 \text{ MPa}$$ It will last more than 10^7 cycles.

FURTHER READING

A. H. Cottrell, *The Mechanical Properties of Matter,* chapters 4 and 8, J. Wiley and Sons, New York, 1964.
H. W. Hayden, W. G. Moffatt, and J. Wulff, *The Structure and Properties of Materials,* vol. III, chapters 1 and 2, J. Wiley and Sons, New York, 1965.
F. A. McClintock and A. S. Argon (eds.), *Mechanical Behavior of Materials,* chapters 1 and 2, Addison-Wesley, Reading, Massachusetts, 1966.
S. P. Timoshenko and D. H. Young, *Elements of Strength of Materials,* D. Van Nostrand Co., Inc., Princeton, N.J., 1962.
L. H. Van Vlack, *Elements of Materials Science,* 3rd ed., chapter 1, Addison-Wesley, Reading, Massachusetts, 1975.
L. H. Van Vlack, *Materials Science for Engineers,* chapters 1–5, Addison-Wesley, Reading, Massachusetts, 1970.

CHAPTER 3

THE STRUCTURE OF SOLIDS

3.1. BONDING BETWEEN ATOMS

All solids are made up of atoms that are held together by the interaction of the outermost (valence) electrons. The valence electrons can move freely in the solid but can only exist in certain stable patterns within the confines of the solid. The nature of the patterns varies according to the ionic, metallic, or covalent bonding. The primary type of bonding can be determined by considering the affinity of atoms for electrons called electronegativity. Atoms like fluorine and oxygen have a high electronegativity because of their propensity to complete their valence shell while inert gases like neon and argon have low electronegativity because their valence shells are already filled.

Electronegativity can be used to determine the nature of bonding in solids, i.e.,

Solid	Electronegativity (eV)
ionic	>2
metallic	<2
covalent	~2

Figure 3-1 shows the electronegativity of some elements. Thus bonding of table salt (NaCl) is ionic, iron (Fe) is metallic, and diamond (C) is covalent.

Although we try to categorize the primary bonding into three major types, these categories are only applicable in limited cases and many materials escape them. For example, silicon atoms share electrons cova-

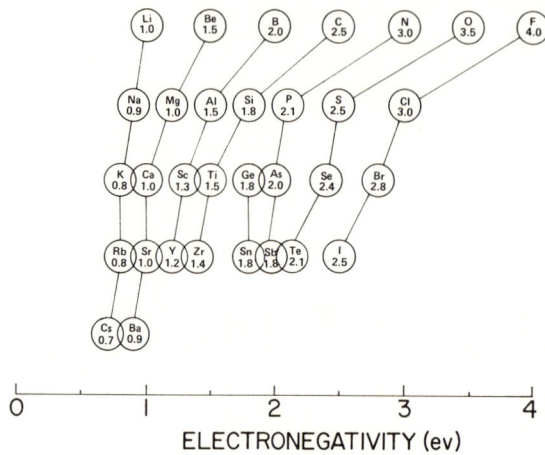

Figure 3-1. Scale of electronegativity of selected elements. The lines connect elements in the same periodic table column. Transition elements are not included (their values vary from 1.6–1.9). (Adapted from L. Pauling, *The Nature of Chemical Bonding*, 3rd edition, p. 93, courtesy Cornell University Press, Ithaca, 1960.)

lently but occasionally a few electrons can be freed and permit limited conductivity (semiconductivity). Therefore, silicon is said to have a limited metallic bonding tendency, as shown in Figure 3-2.

A secondary force between molecules such as in iodine (I_2) is known as a van der Waals force. The iodine molecules can exist as a solid at room temperature but when heated gently, they sublimate destroying the van der Waals bonding. The gaseous units are still covalent iodine (I_2) molecules. The van der Waals forces can be generated by dipoles or polar molecules with asymmetrical distributions of electrical charges.

3.2. ARRANGEMENT OF ATOMS

3.2.1. Atoms of the Same Size

If we assume the atoms can be represented by hard spherical balls (only true at absolute zero), the balls can be packed as shown in Figure 3-3. Note that any cube in the structure can be repeated by moving one lattice spacing, a, in any direction. Thus this type of structure can be represented by a single square. If it is extended into three dimensions it can be represented by a single cube. This is called a simple cubic crystal system. There are six crystal systems beside the simple cubic. The simple

THE STRUCTURE OF SOLIDS 31

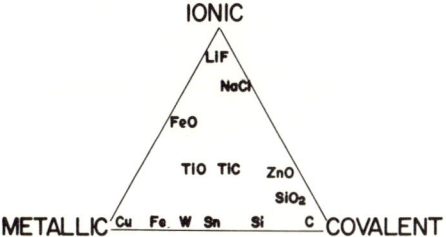

Figure 3-2. Most materials possess a combination of different bonds making a generalization of bonding difficult.

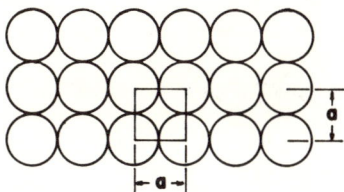

Figure 3-3. Stacking of hard balls (atoms) in simple cubic structure (a is the lattice spacing).

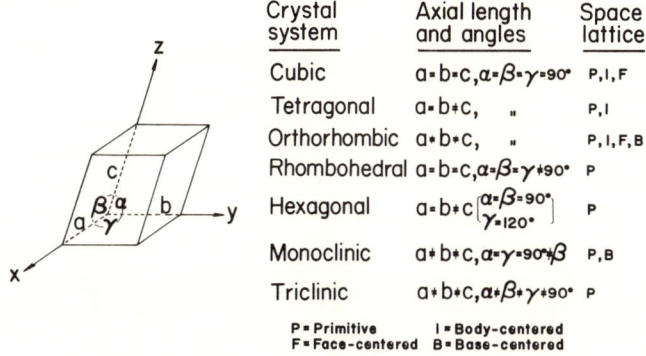

Crystal system	Axial length and angles	Space lattice
Cubic	$a=b=c, \alpha=\beta=\gamma=90°$	P, I, F
Tetragonal	$a=b\neq c$, "	P, I
Orthorhombic	$a\neq b\neq c$, "	P, I, F, B
Rhombohedral	$a=b=c, \alpha=\beta=\gamma\neq 90°$	P
Hexagonal	$a=b\neq c \begin{bmatrix} \alpha=\beta=90° \\ \gamma=120° \end{bmatrix}$	P
Monoclinic	$a\neq b\neq c, \alpha=\gamma=90°\neq\beta$	P, B
Triclinic	$a\neq b\neq c, \alpha\neq\beta\neq\gamma\neq 90°$	P

P = Primitive   I = Body-centered
F = Face-centered   B = Base-centered

Figure 3-4. Crystal structure systems and space lattices.

cubic crystal system is further divided into three space lattices: simple cubic (P), body-centered cubic (I), and face-centered cubic (F). Figure 3-4 shows the space lattices and crystal systems. Any material with a structure represented by one of the 14 space lattices is called crystalline. The cubic and hexagonal systems are the most important for metals and ceramics.

The face-centered cubic (fcc) structure is achieved by stacking the balls as shown in Figure 3-5a. This structure is called close-packed in

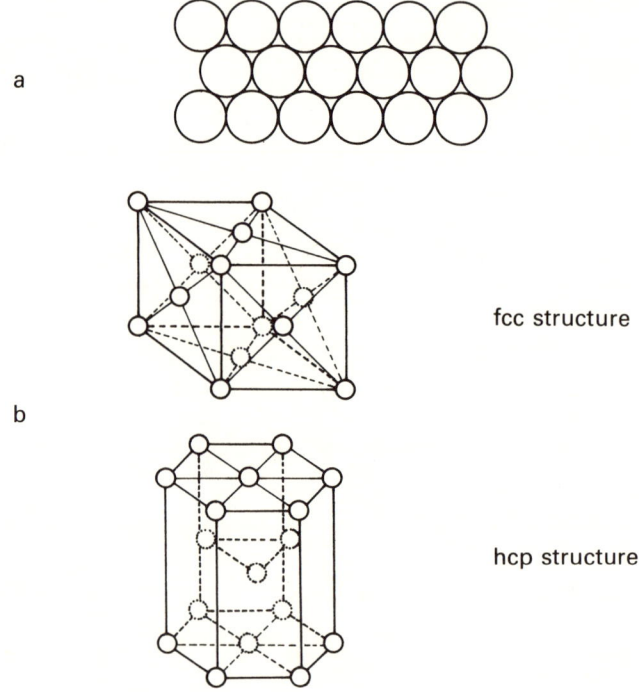

Figure 3-5. Close-packing arrangements: (a) two- and (b) three-dimensional.

three dimensions. Because each atom touches twelve neighbors (hence coordination number is 12), rather than six as in the simple cubic, it results in the most dense atomic volume per unit area. Another close-packed structure is hexagonal as shown in Figure 3-5b. The hexagonal close-packed (*hcp*) structure is characterized by the repeating layers of every other plane—i.e., the atoms in the third layer occupy sites directly over the atoms in the first layer. This can be represented as *ABAB* . . . packing, while the *fcc* structure can be represented by three layers of planes *ABCABC*. . . . Both *fcc* and *hcp* have the highest packing; roughly three-fourths of the unit cell volume is occupied by the atomic volume.

Another common metal structure is the body-centered cubic (*bcc*) in which an atom is situated in the center of the simple cube (Fig. 3-6). Its packing is less efficient; only 68 percent of the cube is accounted for by the atomic volume.

Example 3-1

Chromium has a *bcc* structure with atomic radius of 1.25 Å. Calculate its density (its atomic weight is 52 g).

THE STRUCTURE OF SOLIDS

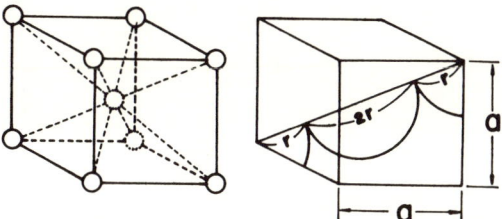

Figure 3.6. Body-centered cubic unit cell. Each unit cell has two atoms and $a = 4r/\sqrt{3}$.

Answer

From Figure 3-6, $a = 4r/\sqrt{3}$ and the density, ρ, is given:

$$\rho = \frac{\text{weight/unit cell}}{\text{volume/unit cell}}$$

$$= \frac{2 \times 52 \text{ g/mol atoms}}{(4 \times 1.25/\sqrt{3})^3 \text{ Å}^3 \; 6.02 \times 10^{23} \text{ atoms/mol}} \quad \text{(2 atoms/unit cell)}$$

$$= 7.18 \text{ g/cm}^3$$

3.2.2. Atoms of Different Size

Most materials used for implants are made of more than two elements except in a few limited applications. When two or more different sizes of atoms are mixed together in a solid, two factors must be considered: (1) the type of site and (2) the number of sites occupied.

Consider the stability of the structure shown in Figure 3-7. In Figure 3-7a and b, because the interstitial atoms touch the larger atoms they are stable, but not in structure c. At some critical value the interstitial atom will fit the space between six atoms (only four atoms are shown in two dimensions), which will give the maximum interaction between atoms and consequently the most stable structure. Thus, at a certain radius ratio of the host and interstitial atoms the arrangement will be the most stable.

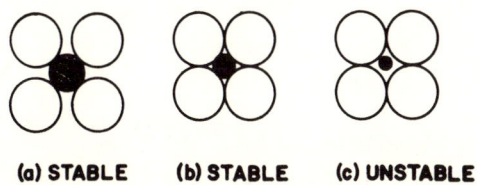

(a) STABLE (b) STABLE (c) UNSTABLE

Figure 3-7. Possible arrangements of interstitial atoms. The critical radius ratio for (b) is given as $r + R = R^{1/2}$; hence $r/R = 0.414$ as given in Figure 3-8.

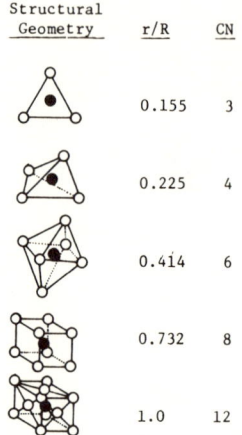

Structural Geometry	r/R	CN
	0.155	3
	0.225	4
	0.414	6
	0.732	8
	1.0	12

Figure 3-8. Minimum radius ratios and coordination numbers.

Figure 3-8 gives the minimum radius ratios for given coordination numbers. Note that these radius ratios are determined solely by the geometric considerations.

Example 3-2

Calculate the minimum radius ratios for coordination number 6.

Answer

For coordination number 6, from Figure 3-8 looking down:

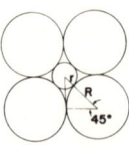

From diagram:

$$\cos 45° = R/(R + r)$$
$$1/\sqrt{2} = R/(R + r)$$
$$\sqrt{2}R = R + r$$
$$r/R = \sqrt{2} - 1 = \underline{0.414}$$

Two-dimensional representation of a structure with coordination number 6.

3.3. IMPERFECTIONS IN STRUCTURES

Imperfections in crystalline solids, sometimes called defects, play a major role in determining the solids' physical properties. Point defects commonly appear as lattice vacancies or as interstitial atoms (Fig. 3-9). The interstitial atoms are sometimes called impurities.

THE STRUCTURE OF SOLIDS

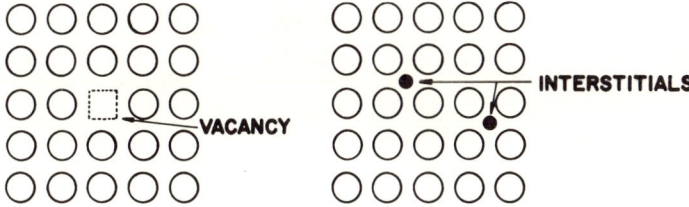

Figure 3-9. Point defects in the form of vacancies and interstitials.

Line defects are created when an extra plane of atoms is displaced or dislocated out of its regular lattice space registry (Fig. 3-10). Planar defects or two-dimensional defects exist at the grain boundaries. Grain boundaries are created when two or more crystals exist in a material. Within each grain all the atoms are in a lattice of one specific orientation. Other grains have the same crystal lattice but different orientations, creating a region of mismatch. The grain boundary is less dense than the bulk; hence most diffusion of gas or liquid takes place along grain boundaries.

Grain boundaries can be seen by polishing and subsequent etching of a "polycrystalline" material. This occurs because the grain boundary atoms possess more energy than the bulk, resulting in a more chemically reactive site at the boundary. Figure 3-11 shows a polished surface of a metal implant. The size of the grains plays an important role in determining physical properties of a material. In general, a fine-grained structure is

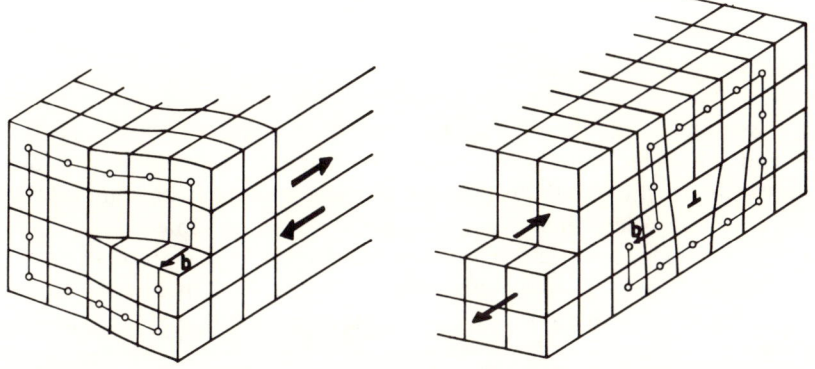

Figure 3-10. Line defects. The displacement is perpendicular to the edge dislocation but parallel to the screw dislocation (left-hand side). The unit length (b) has magnitude and direction; it is called a Burgers vector.

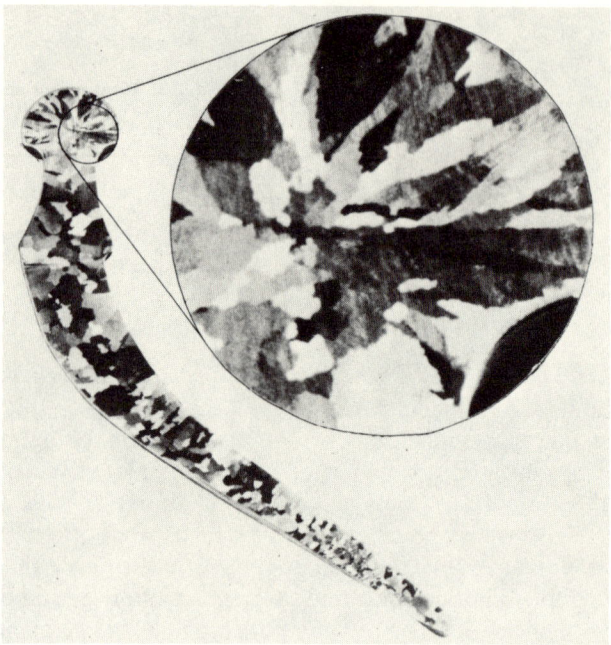

Figure 3-11. Midsection of a femoral component of a hip joint implant which shows grains (Co–Cr alloy). Notice the size distribution along the stem and from the core to the surface.

stronger than the coarser one for a given material because the former contains more grain boundaries which in turn interfere with the movement of atoms during deformation, resulting in a stronger material.

3.4. AMORPHOUS SOLIDS

Some solids like window glass do not have regular crystalline structure and are called amorphous materials. They are usually supercooled from the liquid state so they retain liquidlike structure. Consequently the density is always less than the crystalline state of the same material indicating inclusion of some voids (free volume, Fig. 3-12). Due to the quasiequilibrium state of the structure the amorphous material tends to crystallize. It is also more brittle and less strong than its crystalline counterpart.

It is very difficult to make metals amorphous because the metal atoms are extremely mobile. The ceramics and polymers can be made amorphous because of the relatively sluggish mobility of their molecules.

THE STRUCTURE OF SOLIDS

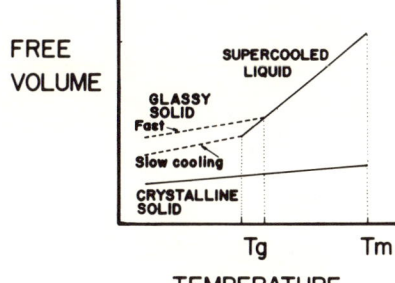

Figure 3-12. Change of volume versus temperature of a solid. The glass transition temperature (Tg) depends on the rate of cooling and below Tg the material behaves as a solid like a window glass.

Example 3-3

Calculate the free volume of 500 g of supercooled iodine from liquid which has a density of 4.8 g/cm³. Assume that the density of amorphous iodine is 4.3 g/cm³ and crystalline density is 4.93 g/cm³.

Answer

The fraction of supercooled iodine is

$$\frac{4.93 - 4.8}{4.93 - 4.3} = 0.21$$

The weight of the supercooled liquid is 0.21 × 500 g = 105 g, therefore the total free volume is

$$\left(\frac{1}{4.3}\frac{cm^3}{g} - \frac{1}{4.93}\frac{cm^3}{g}\right) \times 105 \text{ g} = \underline{3.26 \text{ cm}^3}$$

Upon complete crystallization the volume of the iodine will decrease by 3.26 cm³.

PROBLEMS

3-1. Which of the following compounds have ionic or covalent bonds?

a. Ammonia (NH_3)
b. Salt (NaCl)
c. Carbon tetrachloride (CCl_4)
d. Hydrogen peroxide (H_2O_2)
e. Ozone (O_3)
f. Ethylene ($CH_2\!=\!CH_2$)
g. Water (H_2O)

Answers
Ionic bonding: b.
Covalent bonding: a, c, d, e, f, g.

3-2. Calculate the weight of an iron atom. The density of an atom is 7.87 g/cm³. The Avogadro's number is 6.02 × 10²³. How many iron atoms are contained in 1 cm³?

Answer
For Fe, atomic weight is 55.85 g.

$$\frac{\text{weight}}{\text{atom}} = \frac{55.85 \text{ g/mol}}{6.02 \times 10^{23} \text{ atoms/mol}} = \underline{9.28 \times 10^{-23} \text{ g/atom}}$$

$$\frac{\text{atoms}}{\text{cm}^3} = \frac{6.02 \times 10^{23} \text{ atoms}}{55.85 \frac{\text{g}}{\text{mol}} \text{ mol}} \times \frac{7.87 \text{ g}}{1 \text{ cm}^3} = \underline{8.5 \times 10^{22} \text{ atoms/cm}^3}$$

3-3. Calculate the number of atoms present per cm³ for alumina (Al_2O_3) which have a density of 3.8 g/cm³.

Answer
M.W. of Al_2O_3 = 2 × 26.98 + 16 × 3 = 102 g/mol.

$$\frac{\text{atoms}}{\text{cm}^3} = \frac{3.8 \text{ g} \times 6.02 \times 10^{23} \times 5 \text{ atoms/mol}}{\text{cm}^3 \times 102 \text{ g/mol}} = \underline{1.12 \times 10^{23}}$$

3-4. Calculate the diameters of the smallest cations that can have a 6-fold and 8-fold coordination with O^{-2} ions.

Answer
From Table 5-1 the radius of the oxygen ion is 1.40 Å and from Figure 3-7 the radius ratios for 6-fold and 8-fold coordinations are 0.414 and 0.732 respectively. Therefore

$$\text{6-fold,} \quad 1.4 \times 0.414 \times 2 = \underline{1.16 \text{ Å}}$$

$$\text{8-fold,} \quad 1.4 \times 0.732 \times 2 = \underline{2.05 \text{ Å}}$$

FURTHER READING

A. H. Cottrell, "The Nature of Metals," in *Materials,* ed. D. Flanagan et al., W. H. Freeman & Co., San Francisco, 1967.

THE STRUCTURE OF SOLIDS

R. H. Krock and M. L. Ebner, *Ceramics, Plastics, and Metals,* chapter 3, D. C. Heath Co., Boston, Massachusetts, 1965.

W. G. Moffatt, G. W. Pearsall, and J. Wulff, *The Structure and Properties of Materials,* vol. I, *Structure,* chapters 1–3, J. Wiley and Sons, New York, 1964.

L. Pauling, *The Nature of Chemical Bonding,* 3rd ed., chapter 2, Cornell University Press, Ithaca, New York, 1960.

M. J. Starfield and A. M. Shrager, *Introductory Materials,* McGraw-Hill Book Co., New York, 1972.

CHAPTER 4

METALS AND ALLOYS

The metal atoms are held together by the interactions of the valence electrons (outermost electrons) and the positive metallic ions. The electrons are free to move throughout the solid (sea of electrons) because the valence electrons are not bound between metal ions. Such an arrangement is called the metallic bond, to distinguish it from the covalent bond, in which electrons are shared between one another (Fig. 4-1).

The metallic bonds are not directional because of the diffuse nature of the electrons; this in turn makes the metals able to conduct electricity and heat more easily. Also, the valence electron cloud is responsible for the opaqueness and luster of metals because the electrons interact with light. As the number of valence electrons increases down the periodic table they become more localized and the metallic bonds become more directional, resulting in more brittle metals. This is partially the reason for the high melting temperature of such brittle metals as tungsten and iron.

Metals can be divided into two general categories according to their composition: unalloyed and alloyed. Naturally every atom of unalloyed pure metal has only one kind of neighbor. Because their mechanical, chemical, and thermal properties can be easily improved by alloying metals, most metals used in medicine and dentistry are alloyed.

4.1. ALLOYS AND PHASE DIAGRAMS

When two or more metallic elements are melted and cooled they form either intermetallic compounds or solid solutions and more usually a mixture of each. Such combinations are called alloys. The alloys can be either single or multiphase, depending on temperature and composition. (A *phase* is defined as a physical homogeneous part of a material system.) Thus,

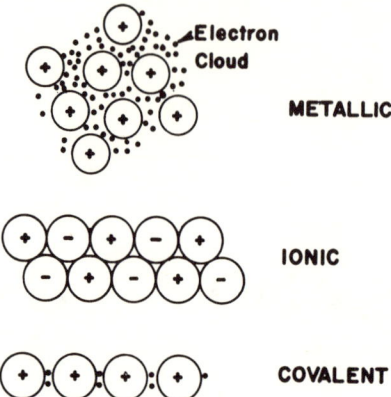

Figure 4-1. Schematic representation of metallic, ionic, and covalent bonding.

liquid and gas are each single phases, but there can be more than one phase for such solids as *fcc* iron and *bcc* iron depending on pressure and temperature. Among multiphase metals, steels are iron-base alloys containing various amounts of a carbide (usually Fe_3C) phase. In this case the carbon atoms occupy the interstitial sites of the iron atoms (cf. Fig. 4-5) and form what is called an interstitial solid solution. Most metal atoms are too large to exist in the interstitial sites. If the two metal atoms are roughly the same size, have the same bonding tendencies, and tend to crystallize in the same types of crystal structure, then a substitutional solid solution may form.

This structure is composed of a random mixture of two different atoms as shown in Figure 4-2. Unless the elements are very similar in properties such a solution will exhibit a limited solubility; i.e., as more substitutional atoms are added in the matrix, the lattice will be more and more distorted until phase separation occurs at solubility limit. In some

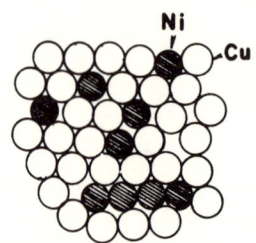

Figure 4.2. Substitutional solid solution of the Cu–Ni system.

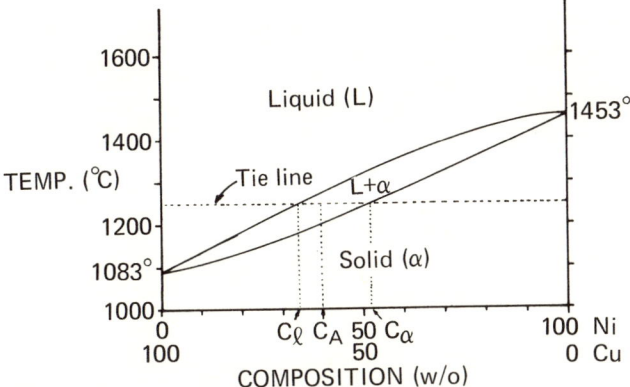

Figure 4-3. The Cu–Ni phase diagram: an example of complete solid solubility. The tie line is explained in the text.

systems complete solid solubility exists, such as in the Cu–Ni system shown in Figure 4-3.

The phase diagram is constructed by first making known compositions of Cu–Ni and then melting and cooling them under thermal equilibrium. During the cooling cycle, we must determine at what temperatures the first solid phase (α) appears and all the liquid disappears. These points will determine the liquidus and solidus line in the phase diagram. From this phase diagram we can determine the types of phase and amount of each element present for a given composition and temperature. Thus if we cool 40 w/o Ni–60 w/o Cu liquid solution (from Fig. 4-3):

Temperature (°C)	Phase (relative amount)	Composition of each phase
above 1270	liquid (all)	40Ni–60Cu
1250	$\begin{cases} \text{liquid (60\%)} \\ \alpha \text{ (40\%)} \end{cases}$	$\begin{cases} \text{33Ni–67Cu} \\ \text{50Ni–50Cu} \end{cases}$
1220	$\begin{cases} \text{liquid (5\%)} \\ \alpha \text{ (95\%)} \end{cases}$	$\begin{cases} \text{26Ni–74Cu} \\ \text{43Ni–57Cu} \end{cases}$
below 1210	α (all)	40Ni–60Cu

The relative amount of each phase present at a given temperature and composition is determined by lever rule after making a horizontal (tie) line at the temperature of interest. Let C_A and C_B be the compositions of element A (Ni) and B (Cu) in the two-phase region met by the tie line (e.g., 1240°C), with the same composition as that given above (40 w/o Ni = C_A); then the amount of element A in the liquid and solid phases can be expressed:

$$\frac{\text{Weight of metal } A}{\text{in the alloy}} = \frac{\text{Weight of metal } A}{\text{in liquid phase}} + \frac{\text{Weight of metal } A}{\text{in solid phase}}$$

$$[W_o \times C_A = W_l \times C_l + W_s \times C_\alpha] \quad (4\text{-}1)$$

where W_o is the weight of the original mixture of metal A and B, and W_l and W_s are the weight of liquid and solid. C_l and C_α are the weight fractions of metal A in the liquid and solid solutions.

Since $W_o = W_l + W_s$ we can substitute this into equation (4-1) and obtain the following expression:

$$\frac{W_s}{W_o} = \frac{C_A - C_l}{C_\alpha - C_l} \quad (4\text{-}2)$$

This can be written another way; that is,

$$\frac{W_s}{W_o} = \frac{\alpha}{\alpha + l} = \frac{C_A - C_l}{C_\alpha - C_l} \left[= \frac{40 - 34}{52 - 34} = 0.33 \right] \quad (4\text{-}3)$$

This principle can be applied in more complicated systems such as Ag–Cu (eutectic) or Fe–C (eutectic + eutectoid) systems as shown in Figures 4-4 and 4-5. The eutectic and eutectoid reactions are defined as:

$$\begin{aligned} L &\rightarrow S_1 + S_2 \\ S_2 &\rightarrow S_1 + S_2 \end{aligned} \quad (4\text{-}4)$$

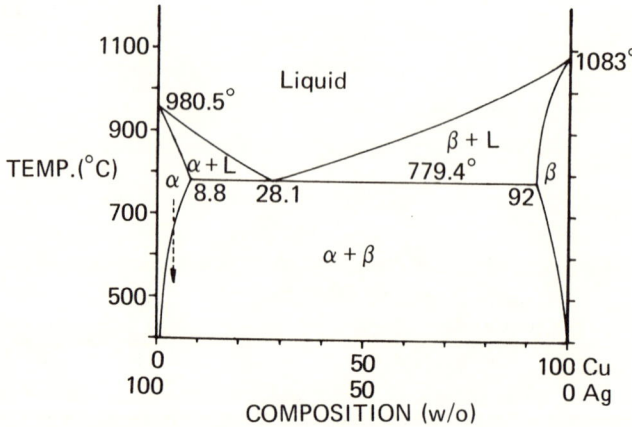

Figure 4-4. The Cu–Ag phase diagram. The dotted vertical line indicates the precipitation hardening by quenching the α phase.

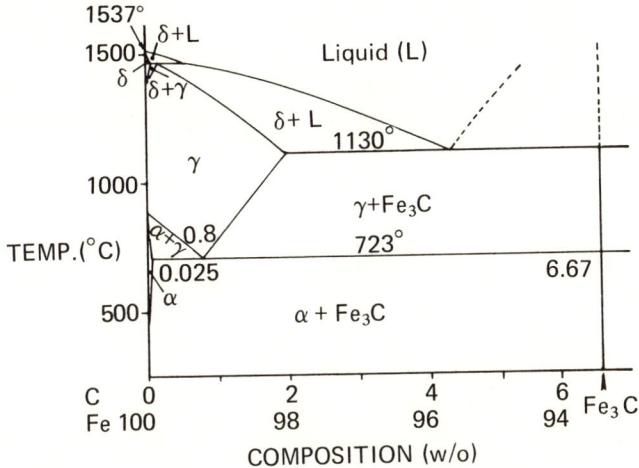

Figure 4-5. The Fe-C phase diagram.

Hence the last liquid will disappear at eutectic temperature and composition (cf. Fig. 4-4).

Example 4-1

Metals of 20 w/o Cu and 80 w/o Ag powders are mixed thoroughly and heated to well above the melting temperature of the alloy. On cooling in a thermodynamic equilibrium condition, calculate the relative amount of Cu and Ag in the phase(s) present:

a. At 1000°C
b. At 780°C
c. At 700°C

Answers

From Figure 4-4.

a. At 1000°C, all liquid (20 w/o Cu + 80 w/o Ag)

b. At 780°C, $\begin{cases} \alpha(8.8 \text{ w/o Cu} + 91.2 \text{ w/o Ag}) \text{ 42 w/o} \\ L(28.1 \text{ w/o Cu} + 71.9 \text{ w/o Ag}) \text{ 58 w/o} \end{cases}$

$$\frac{\alpha}{\alpha + L} = \frac{28.1 - 20}{28.1 - 8.8} = \frac{18.1}{19.3} = \underline{0.94}$$

c. At 700°C, $\begin{cases} \alpha(6 \text{ w/o Cu} + 94 \text{ w/o Ag}) \text{ 88 w/o} \\ \beta(93 \text{ w/o Cu} + 7 \text{ w/o Ag}) \text{ 12 w/o} \end{cases}$

$$\frac{\alpha}{\alpha + \beta} = \frac{93 - 20}{93 - 6} = \frac{73}{87} = \underline{0.88}$$

4.2. IMPERFECTIONS AND STRENGTHENING MECHANISMS

Structural imperfections play a major role in determining the physical properties of a metal, as briefly discussed in the previous chapter. All alloys and multiphase polycrystalline structures have imperfections, although when intentionally created they are not called imperfections.

The interstitial point defects are important in the Fe-C system because the carbon atoms occupy the Fe lattices. The amount of carbon dissolved in iron is limited to 6.67% by weight because the carbon atom is slightly larger than the interstitial sites of the iron lattice (Fig. 4-5). Of course, the substitution solid solution results in alloys that can be strengthened over the original pure metals. A typical property change with composition in Cu–Ni alloys is shown in Figure 4-6. In a two-phase system, increasing the amounts of harder or stronger elements may increase the strength if properly dispersed.

The line defects are important in strain or the work-hardening process because edge and screw dislocations are introduced by the cold work. Cold-work is a measure of deformation below the recrystallization temperature so that the strain-hardening remains in the structure and can be calculated:

$$\text{C.W.} = (A_o - A_f)/A_o \qquad (4\text{-}5)$$

where A_o and A_f are original and final cross-sectional areas, respectively.

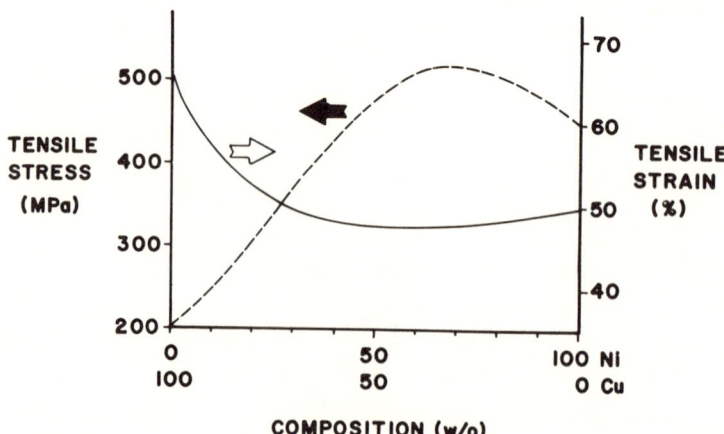

Figure 4-6. Tensile stress and strain changes of copper by adding nickel.

METALS AND ALLOYS

The resultant strain-hardening after cold-work increases hardness, strength, and stiffness, but reduces ductility.

The plane defects are important for recrystallization processes where the microstructure is changed by thermal energy. One of the strengthening processes is the precipitation-hardening of alloys. This is accomplished by rapidly cooling a solid solution of decreasing solubility, as shown in Figure 4-4 by the dotted vertical line. If the quenching is done properly there will be no time for the second phase (β) to form. Hence a quasithermal equilibrium exists but, depending on the amount of thermal energy (i.e., temperature) and time, the second phase (β) will form (precipitation). If the β-phase particles are small and uniformly dispersed throughout the α matrix, their presence can increase the strength greatly. It is important that they be dispersed within an α grain as well as grain boundaries so that the dislocations can be impeded during the deformation process as in the case of cold-working. Carbide (intermetallic compound of metal atoms with carbon, e.g., Fe_3C in Fig. 4-5) precipitates in cobalt-base alloys (e.g., Vitallium®), thus strengthening them.

In relation to the precipitation process, the transformation process is another mechanism of strengthening steel and its alloys. When *fcc* iron or steel is quenched from the austenitic temperature range (γ phase in Fig. 4-5), there is no time for carbon and other alloying elements to form $\alpha + C$ phases. Almost all the carbon atoms must diffuse to form carbide (C) as well as such carbide formers as Cr, Mo, and V, which should concentrate in the carbide, whereas Ni and Si must diffuse into the ferrite (α). These reactions take time at low temperatures (below 400°C). Since the *fcc* structure of austenite is not in equilibrium a driving force develops and at low enough temperatures this driving force becomes sufficient to force transformation by shear. The resulting structure is a tetragonal martensite instead of the body-centered ferrite. The martensite is extremely hard because it is noncubic (has less slip combinations than the *bcc* structure) and the interstitially entrapped carbon prevents slipping. Martensite is the hardest iron-rich phase material but is extremely brittle. Hence tempering by heating (600°C) is necessary to make it tough and strong.

$$\text{Martensite} \xrightarrow{\text{Tempering}} \alpha + \text{carbide } (C; Fe_3C) \tag{4-6}$$

Example 4-2

Titanium changes its structure from *hcp* structure ($a = 2.956$ Å, $c = 4.683$ Å) to *bcc* structure ($a = 3.32$ Å) above 880°C. Calculate the volume change in cm³/g by heating above the transition temperature. The density is 4.54 g/cm³ with an atomic weight of 47.9 g.

Answer

Area of hexagon

$$= a \times \frac{\sqrt{3}}{2} a \times \frac{1}{2} \times 6$$

$$= \frac{3\sqrt{3}}{2} a^2 \text{ Å}^2$$

$V_{hcp} = \dfrac{3\sqrt{3}}{2} a^2 c = 106.3$ Å^3 which contains 6 atoms

$V_{bcc} = a^3 = 36.594$ Å^3 which contains 2 atoms

hcp: $\dfrac{106.3 \text{ Å}^3}{6 \text{ atoms}} = \dfrac{106.3 \text{ Å}^3 \times 6.02 \times 10^{23} \text{ atoms/mol}}{6 \times 47.9 \text{ g/mol atoms}} = \underline{0.2227 \text{ cm}^3/\text{g}}$

bcc: $\dfrac{36.594 \text{ Å}^3}{2 \text{ atoms}} = \dfrac{36.594 \text{ Å}^3 \times 6.02 \times 10^{23} \text{ atoms/mol}}{2 \times 47.9 \text{ g/mol atoms}} = \underline{0.23 \text{ cm}^3/\text{g}}$

The difference is $\underline{0.0073 \text{ cm}^3/\text{g}}$.

4.3 CORROSION OF METALS

As mentioned previously, because the valence electrons of metals are not strongly bound to the nucleus, they can be removed easily and become

$$M \rightarrow M^{+n} + ne^- \qquad (4\text{-}7)$$

where n is the number of valence electrons. If the electrons are removed from the system then the equilibrium is disturbed and the reaction will be driven to the right. The result is a corrosion cell. To be a continuous corrosion process the system must have two reactions, i.e., one producing electrons and the other consuming them. Thus, the electron transfer must occur for the corrosion to take place. The pole-producing electrons are called the anode (being oxidized), while the consuming electrons are called the cathode (being reduced). Cathode reduction can occur at any metal surface capable of transferring electrons to the anode. It can be the same metal portion or different phases of the same metal.

Corrosion occurs at the point where the electrons are produced (anode) and the metal ions generated dissolve into solution or combine with other species in the environment.

Most corrosion occurs through the interaction of the solution and oxidation process. The dissolved ions and electrons [equation (4-7)] will build up an electrical potential, called an electrode potential, that depends

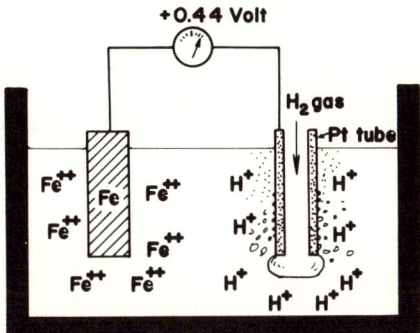

Figure 4-7. Iron–hydrogen corrosion cell. The platinum tube is porous.

on the nature of the metal and the solution. The electrode potential of any material is measured by using a standard hydrogen electrode (Fig. 4-7). The equilibrium of hydrogen reaction can be written

$$H_2 \rightarrow 2H^+ + 2e^- \tag{4-8}$$

The electrode potentials of various metals are given in Table 4-1. If two dissimilar metals are electrically connected in a medium containing their ions, corrosion will occur. The noble metal will be the cathode and the base metal the anode, according to the electrode potentials shown in Table 4-1. This type of corrosion takes place readily on a microscopic level as shown in Figure 4-8. In both cases a microgalvanic cell is set up due to the electrode potential differences; the ferrite is less noble than the carbide; and the grain boundary is the region of disorder and thus of higher reactivity, therefore less noble than the interior of the grains.

Local differences in concentration of electrolyte and oxygen may also lead to corrosion. In the region of high oxygen concentration of an aqueous solution the following reaction will occur to the right by consuming electrons:

$$2H_2O + O_2 + 4e^- \rightarrow 4(OH)^- \tag{4-9}$$

Regions of lower oxygen concentration become anodic to provide those electrons, as shown in Figure 4-9.

Any region of distortion or stress will be anodic with respect to the lesser distorted or stressed region of the same material because the stressed region has a higher energy level than the nonstressed region. This is the reason why the head or bent portion of a nail is more readily

Table 4-1. Electrode Potentials of Various Ions

Ions	Potential (volts)	
Li^+	+2.96	Anode
K^+	+2.92	
Ca^{2+}	+2.90	
Na^+	+2.71	
Mg^{2+}	+2.40	
Ti^{3+}	+2.0	
Al^{3+}	+1.70	
Zn^{2+}	+0.76	
Cr^{2+}	+0.56	
Fe^{2+}	+0.44	
Ni^{2+}	+0.23	
Sn^{2+}	+0.14	
Pb^{2+}	+0.12	
Fe^{3+}	+0.045	
H^+	0.000	Reference
Cu^{2+}	−0.34	
Cu^+	−0.47	
Ag^+	−0.80	
Pt^{4+}	−0.86	
Au^+	−1.50	Cathode

corroded. This is also the reason why manufacturers advise surgeons not to bend or twist the devices unless absolutely necessary.

Corrosion can be accelerated in the presence of external stress. This is called stress or fatigue corrosion and predominates under repeated loading condition. Most implants are subject to this type of corrosion. The endurance (fatigue) limit of a metal can be lowered by this process.

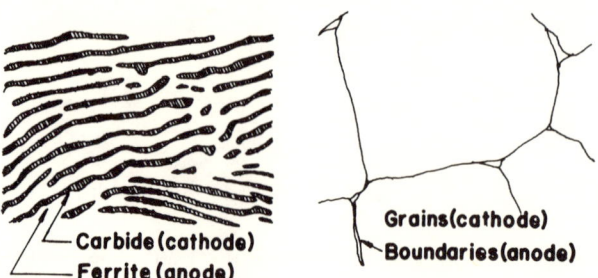

Figure 4-8. Galvanic corrosion caused by energy difference in microstructures. Left: Galvanic microcell of pearlite. Right: Galvanic microcell of grain and grain boundary.

METALS AND ALLOYS

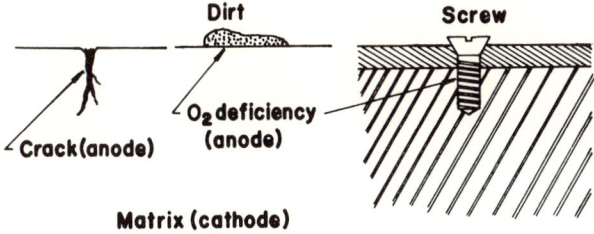

Figure 4-9. Example of oxygen-deficient corrosion cells.

A brief summary of various galvanic corrosion cells is given in Table 4-2. The transfer of electrons occurs because of differences in composition, energy level, and electrolytic environment. The electrode with higher potential or energy is the anode that undergoes corrosion.

If a metal forms a coherent, stable oxide film of protective surface then the metal is said to be passivated. Such an oxide film will slow the dissolution of metal ions into the environment, thus retarding the corrosion. This is the reason titanium and chromium are corrosion-resistant despite the relative positions of their electrode potentials (cf. Table 4-1). Aluminum also forms a stable oxide film but it becomes unstable in NaCl solutions. Thus, the corrosion resistance of metals in salt water is different and depends on their passivity. These values are given in Table 4-3. If the passivation film breaks down corrosion will take place at that point, which becomes anodic and the rest of the material becomes cathodic. Because the broken-down anodic region should provide electrons for all the cathodic regions it will result in an accelerated corrosion, or pitting corrosion, at the anode.

There are two reasons why the corrosion of metals or any deterioration of implant materials is important; one is obviously the weakening

Table 4-2. Summary of Galvanic Corrosion

Differences	Examples	
	Anode	Cathode
Composition	Fe	H_2
	Ferrite (α)	Carbide
Energy level	Boundaries	Grain
	Stressed region	Nonstressed region
Electrolytic	Low PO_2	High PO_2
Environment	Dilute solution	Concentrated solution

Table 4-3. The Galvanic Series for Metals and Alloys, Including Some Stainless Steels[a]

Anodic end (electropositive)

	Magnesium	
	Magnesium alloys	
	Zinc	
	Aluminum	
	Cadmium	
	Aluminum alloy	
	Carbon steel	
	Copper steel	
	Cast iron	
	4 to 6% Cr steel	
A	{ 12 to 14% Cr steel 16 to 18% Cr steel 23 to 30% Cr steel }	Active
B	{ 7% Ni, 17% Cr steel 8% Ni, 18% Cr steel 14% Ni, 23% Cr steel 20% Ni, 25% Cr steel 12% Ni, 18% Cr, 3% Mo steel }	Active
	Lead–tin solder	
	Lead	
	Tin	
	Nickel	
C	{ 60% Ni, 15% Cr, 20% Fe Inconel 80% Ni, 20% Cr }	Active
	Brasses	
	Copper	
	Bronzes	
	Nickel–silver (Ni-rich brass)	
	Copper–nickel	
	Monel metal	
	Nickel	
C	{ 60% Ni, 15% Cr, 20% Fe Inconel 80% Ni, 20% Cr }	Passive
A & B	{ 12 to 14% Cr steel 16 to 18% Cr steel 7% Ni, 17% Cr steel 8% Ni, 18% Cr steel 14% Ni, 23% Cr steel 23 to 30% Cr steel 20% Ni, 25% Cr steel 12% Ni, 18% Cr, 3% Mo, steel }	Passive
	Silver	
	Graphite	

Cathodic end (electronegative)

[a] Note that the role of passivation in moving the stainless steels toward the bottom of the table. (Adapted from *Stainless Steels* by C. A. Zapffe, American Society for Metals, 1949, by permission from the publisher.)

METALS AND ALLOYS

effect and the other is the effect of the corrosion by-products on the tissues. A low corrosion rate of surgical stainless steel (316 and 316L types) of about 0.2 μm/year or less is not detrimental to the mechanical properties of an implant. However, if we calculate the number of ions released in the area of a pinhead (0.01 cm²) by using the corrosion rate of 0.2 μm/year of the stainless steel (density, 9 g/cm³), we find that

$$\text{C.R.} = 0.2 \frac{\mu m}{\text{year}} = \frac{0.2 \times 10^{-4} \text{ cm}}{365 \times 24 \times 3600 \text{ s}} = 6 \times 10^{-13} \text{ cm/s}$$

For an area of 0.01 cm² the number of ions released per second can be calculated:

$$\left(\frac{6 \times 10^{-13} \text{ cm}}{\text{s}}\right) \cdot (0.01 \text{ cm}^2) \cdot \left(\frac{9 \text{ g}}{\text{cm}^3}\right) \cdot \left(\frac{1 \text{ mol}}{60 \text{ g}}\right) \cdot \left(\frac{6 \times 10^{23} \text{ atoms}}{\text{mol}}\right)$$
$$= 5.4 \times 10^8 \frac{\text{ions}}{\text{s}}$$

This is over 500 million ions per second released from an area about the size of a pinhead (1 × 1 mm). This amount of ions may not be significant to irritate tissues but may have significance in long-term implantation.

Example 4-3

Calculate the corrosion rate in mm/year of a platinum anode through which 10 μA/cm² current flows.

Answer

Assuming a uniform current flow throughout the anode,

$$\frac{(10 \times 10^{-6} \text{ A/cm}^2)(3.5 \times 10^7 \text{ s/year})}{1.6 \times 10^{-19} \text{ C/electron}} = 2.19 \times 10^{21} \text{ electrons/cm}^2 \cdot \text{year}$$
$$= 5.47 \times 10^{20} \text{ Pt atoms/cm}^2 \cdot \text{year}$$

$$\frac{(5.47 \times 10^{20} \text{ Pt atoms/cm}^2 \cdot \text{year})(195 \text{ g/mol})}{(6.02 \times 10^{23} \text{ Pt atoms/mol})(21.45 \text{ g/cm}^3)} = 0.0083 \text{ mm/year } (8.3 \text{ }\mu\text{m/year})$$

This corrosion rate is far more than the 0.2 μm/year cited previously although platinum is a noble metal. This rate occurs because electrical energy input allows the corrosion to take place at a faster rate.

CHAPTER 4

PROBLEMS

4-1. A 1080 plain steel (0.8 w/o C) is cooled from liquid.

a. What phases are present at 1000°C and 720°C?
b. What are the compositions of each phase?
c. What is the relative amount of each phase?

Answers
From Figure 4-5,

a.
$$1000°C: \text{all } \gamma$$
$$720°C: \alpha + Fe_3C$$

b.
$$\gamma: 0.8 \text{ w/o C} + 99.2 \text{ w/o Fe}$$
$$\begin{cases} \alpha: 0.025 \text{ w/o C} + 99.975 \text{ w/o Fe} \\ Fe_3C: 6.67 \text{ w/o C} + 93.33 \text{ w/o Fe} \end{cases}$$

c.
$$1000°C: \text{all } \gamma$$
$$720°C: \begin{cases} \alpha: 88\% \\ Fe_3C: 12\% \end{cases}$$

4-2. The following list gives Poisson's ratios for some metals.

a. List them in order of most volume changes to least.
b. Determine Poisson's ratios for various tissues and compare them with the list.

Material	Poisson's ratio
Tungsten	0.27
Iron and steel	0.28
Aluminum	0.34
Copper	0.35
Lead	0.4

Answer
The smaller Poisson's ratio the larger the volume change by deformation, therefore lead undergoes the least volume change.

4-3. A precisely machined 316 stainless steel rod has a diameter of 2.000 cm. After tensioning with a stress of 300 MPa:

a. Determine the strain.
b. Calculate the cross-sectional area change.

Answers
a.
$$\text{Area} = \pi r^2 = \pi(1.000 \text{ cm})^2 = 3.14 \text{ cm}^2$$
$$\text{Strain} = \frac{\sigma}{E} = \frac{300 \text{ MPa}}{200 \text{ GPa}} = 1.5 \times 10^{-3} \text{ cm/cm}$$

b.
$$\epsilon_x = -\nu \times \epsilon_z = -0.28 \times 1.5 \times 10^{-3} = -4.2 \times 10^{-4} \text{ cm/cm}$$
$$\text{Final diameter} = 2.000 - (2.000)(4.2 \times 10^{-4}) = 2 \times 0.99958$$

Therefore,
$$\Delta A = \pi[(1.000)^2 - (0.99958)^2]$$

METALS AND ALLOYS

$$\Delta A = \underline{0.00264 \text{ cm}^2} \; (\sim 0.08\%)$$

4-4. An iron sheet was cold-rolled from 4 mm to 3 mm thick with negligible change in width. Calculate the amount of cold work done on the sheet.

Answer

$$\text{C.W.} = \frac{A_o - A_f}{A_o}$$

since $A_o = W \cdot 4$ and $A_f = W \cdot 3$

$$\text{C.W.} = \frac{W \cdot 4 - W \cdot 3}{W \cdot 4}$$

$$= \underline{0.25 \; (25\%)}$$

4-5. a. Which of the following binary solids can be precipitation-hardened? Ni–Cu, Cu–Ag, Fe–C.
b. Give the maximum amount of other elements that can be present in (a) and their temperatures.

Answers
a. Cu–Ag and Fe–C.
b. 8.8 w/o Cu for α phase $\Big\}$ Cu–Ag at 779.4°C
   92 w/o Cu for β phase
   0.025 w/o C for Fe–C at 723°C

4-6. Calculate the amount of the volume change when iron is oxidized to FeO ($\rho = 5.95$ g/cm³) and when Ca is changed to CaO ($\rho = 3.4$ g/cm³).

Answers

$$\rho_{Fe} = 7.87 \text{ g/cm}^3, \; \text{M.W.} = 55.85 \text{ g/mol}$$
$$\frac{55.85 \text{ g/mol}}{7.87 \text{ g/cm}^3} = 7.1 \text{ cm}^3/\text{mol}$$
$$\rho_{FeO} = 5.95 \text{ g/cm}^3, \; \text{M.W.} = 71.85 \text{ g/mol}$$
$$\frac{71.85 \text{ g/mol}}{5.95 \text{ g/cm}^3} = 12.08 \text{ cm}^3/\text{mol}$$

$$\text{FeO:} \; \Delta V = \frac{12.08 - 7.1}{7.1} = \underline{0.7} \; (70\% \text{ volume increase by oxidation})$$

$$\rho_{Ca} = 1.55 \text{ g/cm}^3, \; \text{M.W.} = 40 \text{ g/mol}$$
$$\frac{40 \text{ g/mol}}{1.55 \text{ g/cm}^3} = 25.8 \text{ cm}^3/\text{mol}$$

$$\rho_{CaO} = 3.4 \text{ g/cm}^3, \text{ M.W.} = 56 \text{ g/mol}$$
$$\frac{56 \text{ g/mol}}{3.4 \text{ g/mol}} = 16.47 \text{ cm}^3/\text{mol}$$

$$\text{CaO: } \Delta V = \frac{16.47 - 25.8}{25.8} = \underline{-0.36 \text{ (36\% volume decrease)}}$$

4-7. A man broke his forearm and an orthopedic surgeon decided to use an internal fixation procedure. If he uses an intramedullary rod that can be approximated as a circular tube with inside and outside diameters of 0.8 cm and 1.0 cm and 15 cm in length, how many ions are released from the inside surface in 3 months? Assume a corrosion rate of 0.2 μm/yr and that the rod is made from stainless steel ($\rho = 9$ g/cm^3).

Answer

$$\text{Surface area} = \pi DL = \pi(0.8)(15) \text{ cm}^2$$
$$\text{Corrosion rate} = \frac{0.2 \ \mu\text{m}}{\text{year}} = \frac{0.2 \times 10^{-4} \text{ cm}}{\text{year}}$$

$$\frac{\text{ions}}{\text{month}} = \frac{\pi \times 0.8 \times 15 \text{ cm}^2 \times 9 \text{ g} \times 6.02 \times 10^{23} \text{ ions} \times \text{mol} \times 0.2 \times 10^{-4} \text{ cm year}}{\text{cm}^3 \times \text{mol} \times 60 \text{ g} \times \text{year} \times 12 \text{ months}}$$
$$= 0.57 \times 10^{19} \text{ ions/month}$$

Therefore, for 3 months the number of ions released will be

$$\underline{1.7 \times 10^{19} \text{ ions}}$$

4-8. Two pieces of metal, one platinum and the other copper, are immersed in water and connected by copper wires.

a. Which metal becomes cathode?
b. What is the maximum electrode potential that can be developed?

Answer
From Table 4-2 the electrode potentials are

$$\text{Pt: } -0.86 \text{ V (cathode)}$$
$$\text{Cu: } -0.34 \text{ V (anode)}$$
$$\underline{-0.52 \text{ volt maximum potential}}$$

METALS AND ALLOYS

4-9. The dental amalgam is an alloy of silver and tin as shown in the following figure for their phase diagram.

a. Calculate the theoretical weight percent of the silver and tin of the γ phase (Ag_3Sn).
b. Fill in the missing phases of (a) to (f).

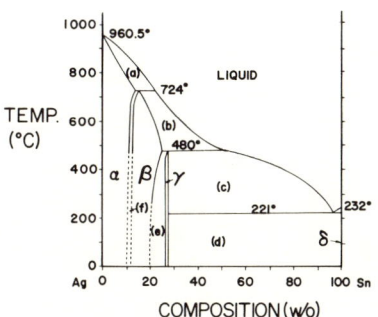

Phase diagram of Ag–Sn.

Answers
a. Since the molecular weights of silver and tin are 107.87 and 118.69 g/mol, respectively,

$$\frac{Sn}{Ag + Sn} = \frac{118.69}{107.87 \times 3 + 118.69} = 0.2684 (26.84 \text{ w/o, tin})$$

This is the weight percent of tin in the Ag–Sn amalgam (γ phase).

b. (a) $\alpha + L$ (b) $\beta + L$ (c) $\gamma + L$ (d) $\gamma + \delta$ (e) $\beta + \gamma$ (f) $\alpha + \beta$

FURTHER READING

L. V. Azàroff, *Introduction to Solids*, chapters 4 and 5, McGraw-Hill Book Co., New York, 1960.
M. G. Fontana and N. O. Greene, *Corrosion Engineering*, McGraw-Hill Book Co., New York, 1967.
A. G. Guy, *Essentials of Materials Science*, chapter 2, McGraw-Hill Book Co., New York, 1976.
A. G. Guy, *Physical Metallurgy for Engineers*, Addison-Wesley, Reading, Massachusetts, 1962.
L. H. Van Vlack, *A Textbook of Materials Technology*, chapters 3–6, Addison-Wesley, Reading, Massachusetts, 1973.
L. H. Van Vlack, *Materials Science for Engineers*, chapters 6 and 22, Addison-Wesley, Reading, Massachusetts, 1970.

CHAPTER 5

CERAMIC MATERIALS

Ceramics contain metallic and nonmetallic elements that are mostly bonded ionically or covalently. As noted in Chpater 3, because their bonds lack free electrons ceramics are poor conductors of electricity and heat. Lack of free electrons makes them also transparent to light. The ionic bonds are highly directional and stable; therefore they have relatively higher melting temperatures, on the average, than metals or polymers. Generally they are also harder and more resistant to chemical changes. Other factors influencing the structure and property relationship of the ceramic materials are radius ratio (Section 3.2) and relative electronegativity between the positive and negative ions, although the net electrical charge of any material should be zero.

Recently, ceramic materials have been given much attention as candidates for implant materials because they possess some highly desirable characteristics for some applications. Ceramics have been used for some time in dentistry as crowns because of their inertness to body fluids, high compressive strength, and good aesthetic appearance.

Carbon materials have been used as artificial heart valve disks, percutaneous buttons and leads, and dental implants. Although their black color is a drawback, they have other desirable qualities, such as good biocompatibilities and ease of fabrication.

5.1. ATOMIC BONDING AND ARRANGEMENT

When (neutral) atoms such as sodium (metal) and chloride (nonmetal) are ionized, sodium will lose an electron and chloride will gain an

Table 5-1. Atomic and Ionic Radii of Some Elements (Units Are Å)[a]

	Group I			Group II			Group VI			Group VII	
Element	Atomic radius[b]	Ionic radius	Element	Atomic radius[b]	Ionic radius	Element	Atomic radius[b]	Ionic radius	Element	Atomic radius[b]	Ionic radius
Li^+	1.52	0.68	Be^{2+}	1.11	0.31	O^{2-}	0.74	1.40	F^-	0.71	1.36
Na^+	1.86	0.95	Mg^{2+}	1.60	0.65	S^{2-}	1.02	1.84	Cl^-	0.99	1.81
K^+	2.27	1.33	Ca^{2+}	1.97	0.99	Se^{2-}	1.16	1.98	Br^-	1.14	1.95

[a] Taken from M. J. Starfield and M. A. Shrager, *Introductory Materials Science*, p. 64, McGraw-Hill Book Co., N.Y. 1972, by permission from the publisher.
[b] Covalent.

electron,

$$Na \rightarrow Na^+ + e^-$$
$$e^- + Cl \rightarrow Cl^- \quad (5\text{-}1)$$

Thus sodium and chloride can make an ionic compound by the strong affinity of the positive and negative ions. This is usually done in an aqueous solution. The negatively charged ions are much larger than the positive ions due to the gain and loss of electrons as given in Table 5-1. The radius of an ion varies according to the coordination numbers: the higher the coordination number the larger the radius. For example, oxygen ion (O^{2-}) has a radius of 1.28, 1.40, and 1.44 Å for coordination numbers 4, 6, and 8, respectively.

Ceramics can be classified according to their structural compounds, of which $A_m X_n$ is an example. The A is for metal and X is for nonmetal elements, m and n are integers. The simplest case of this system is the AX structure, which has three types, as shown in Figure 5-1. The difference between these structures is caused by the relative size of the ions (radius

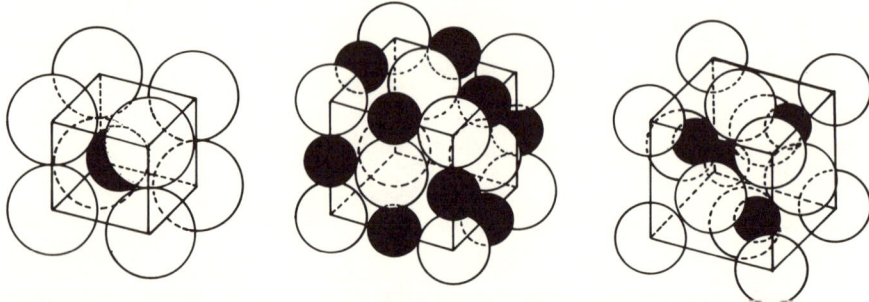

Figure 5-1. AX structures of ceramics. The dark spheres represent positive ions (A^+) and the circled ones represent negative ions (X^-). (a) CsCl; (b) NaCl; (c) ZnS.

CERAMIC MATERIALS

Table 5-2. Selected A_mX_n Structures

Prototype compound	Lattice of A (or X)	Coordination number of A (or X) sites	Sites filled	Minimum r_A/R_x	Other compounds
CsCl	Simple cubic	8	All	0.732	CsI
NaCl	fcc	6	All	0.414	MgO, MnS, LiF
ZnS	fcc	4	½	0.225	β-SiC, CdS, AlP
Al₂O₃	hcp	6	⅔	0.414	Cr₂O₃, Fe₂O₃

ratios). If the positive and negative ions are about the same size ($r_A/R_x > 0.732$), the structure becomes a simple cubic (CsCl structure). The face-centered cubic structure arises if the relative size of the ions are quite different since the positive ions can be fitted in the tetragonal or octagonal spaces created among larger negative ions. These are summarized in Table 5-2. The aluminum and chromium oxide belong to the A_2X_3 type of structure. The O^{2-} ions form close-packing hexagonals while the positive ions (Al^{3+}, Cr^{3+}) fill in two-thirds of the octahedral sites leaving one-third vacant.

Another important structure group is the silicones, which are made of tetrahedral chains as shown in Figure 5-2. This tetrahedral structure is also characteristic of silicates, which can be of the sheet or network type. Network-type silicates exist in many forms, including quartz (crystal form) and fused silica (glassy form).

Piezoelectric ceramics have been studied closely because tissues show similar electromechanical properties. One of the piezoelectric ceramics, barium titanate ($BaTiO_3$), is given in Figure 5-3a. This type of material can be elongated or shortened in an electric field because it has electrical dipoles created by the slight displacement of the relative positions of the negative and positive ions. For example, a positive titanium ion, Ti^{4+} of barium titanate has a choice of two locations, as shown in Figure 5-3b. Because neither is at the center of the unit cell, the centers of positive and negative charges are not coincidental, resulting in an electrical dipole.

The piezoelectric phenomenon is a reversible process and can be expressed simply:

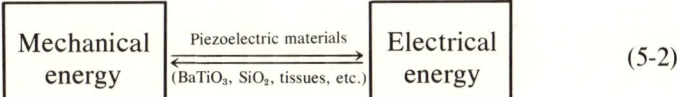

(5-2)

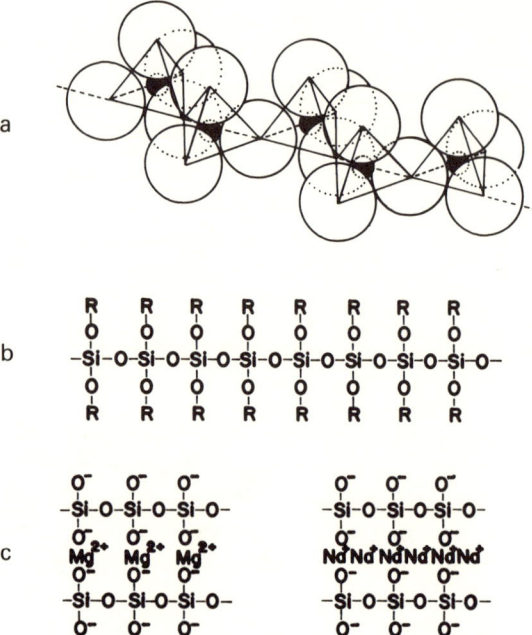

Figure 5-2. Tetrahedral chains: (a) basic (SiO_3) chain; (b) siloxane (R = H, OH, CH_3, C_2H_5, -⟨O⟩, etc.); (c) silicates. (Redrawn after L. H. Van Vlack, *A Textbook of Materials Technology*, p. 221, Addison-Wesley, Reading, Mass., 1973, by permission from the publisher.)

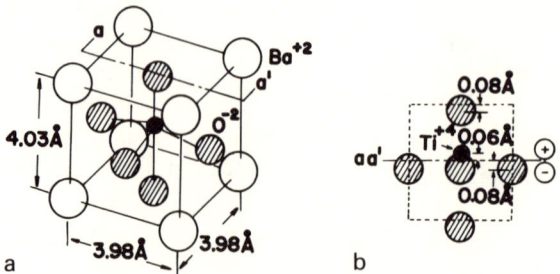

Figure 5-3. (a) Barium titanate unit cell structure and (b) formation of dipoles due to the eccentric location of Ti^{4+} ions in the unit cell structure.

5.2. PHYSICAL PROPERTIES

Ceramics are generally hard; in fact, one measure of hardness in common use is calibration against ceramic materials. On this scale, diamond is the hardest, Moh's scale 10, and talc ($Mg_3Si_4O_{10}(OH)_2$) is the softest (scale 1); others, such as alumina (Al_2O_3; 9), quartz (SiO_2; 8), and apatite ($Ca_5P_3O_{12}F$; 5), are in between. Another characteristic of ceramic materials is their high melting temperature which is due to their high ionic bond energy.

Unlike metals and polymers, ceramics are hard to shear, owing to the ionic nature of bonding, as shown in Figure 5-4. In order to shear, the plane of atoms should slip past each other. However, for ceramic materials ions with the same electric charge repel each other; hence moving the plane of atoms is very difficult because slip is resisted by the repelling forces. This makes ceramics brittle and hard to creep at room temperature. Ceramics are also very sensitive to notch or micro cracks because of stress concentration around the tips of cracks.

Stress concentration by a crack of length c was formulated by Griffith [*Phil. Trans. Roy Soc. (London), A221,* 168, 1921]:

$$\sigma_c/\sigma = 2\sqrt{c/r} \qquad (5\text{-}3)$$

This indicates that even though the applied stress (σ) is small, the concentrated stress (σ_c) can become very large, because the radius of a

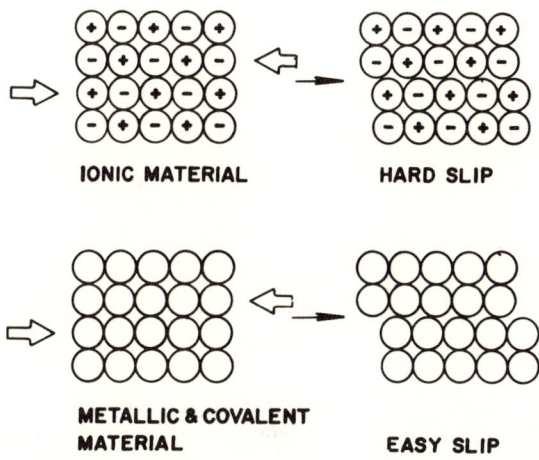

Figure 5-4. Schematic two-dimensional illustration of slips in ionic and nonionic bonding materials.

crack (r) can be as small as the atomic spacing—on the order of 1 Å. It is believed that the stress concentration factor (σ_c/σ) can become 100 or even 1000, which is why the tensile strength of brittle materials is much lower than their theoretical value. This is also why the tensile strength of ceramics is lower than their compressive strength. In compression any cracks or pores can be closed, but in tension the cracks act as stress concentrators, as shown in Figure 5-5. In fact, if a ceramic is made flaw-free, it becomes very strong, even in tension. Glass fibers made this way have tensile strengths twice that of steel (~7 GPa). However, the strength is somewhat unpredictable, as shown in Table 5-3.

It is interesting to note that when cracks are introduced as pores, relative strength is only a function of the pore volume fraction (Fig. 5-6), regardless of the type of material. This suggests that the stress concentration is a predominant cause of failure regardless of the type of bonding (or material).

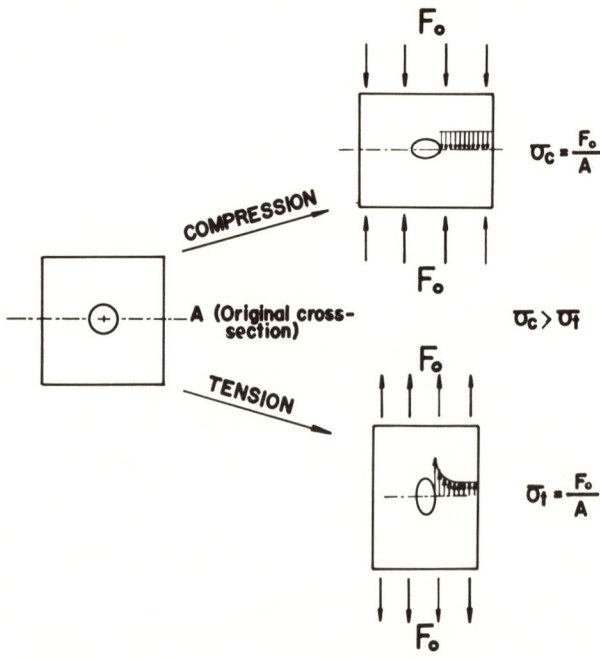

Figure 5-5. Effect of a crack inside a material when the material is subjected to compression and tension. The crack does not grow with compression while the opposite is true with tension because of the stress concentration (indicated with small arrows), provided the material does not undergo plastic deformation. Thus brittle materials such as ceramics have higher strength under compression than under tension.

CERAMIC MATERIALS

Table 5-3. Mechanical Properties of Some Ceramics[a]

Ceramic	Modulus of elasticity (GPa)	Compressive strength (MPa)
Al_2O_3 crystals	379	345–1034
Sintered Al_2O_3 (5% porosity)	365	207–345
Silica glass	72	107
Pyrex glass	69	69

[a] Taken from W. D. Kingery, *Introduction to Ceramics*, pp. 599 and 610, J. Wiley and Sons, New York, 1960, by permission from the publisher.

5.3. DETERIORATION OF CERAMIC MATERIALS

Although the deterioration of ceramics rarely causes any problems in ordinary applications (such as concrete and brick), when they are used as an implant *in vivo* their strength decreases. The polycrystalline ceramics such as alumina and calcium aluminate ($CaO \cdot nAl_2O_3$) show deterioration even in static loading conditions (with a known weight on the sample) for both porous and solid samples, as shown in Figure 5-7. The deterioration is thought to be mainly caused by the attack of water molecules on grain boundaries that have higher energy than the bulk material. Another reason is that the micro cracks or flaws propagate slowly under load until they are large enough to cause a catastrophic failure.

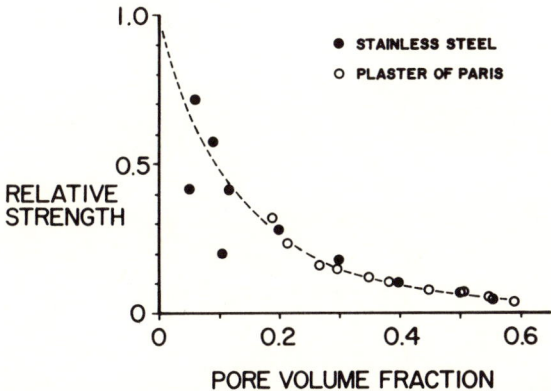

Figure 5-6. Effect of porosity on the tensile strength on stainless steel and plaster of paris. The same is true with polymers. (Redrawn from W. D. Kingery, *Introduction to Ceramics*, p. 622, J. Wiley and Sons, New York, 1960.)

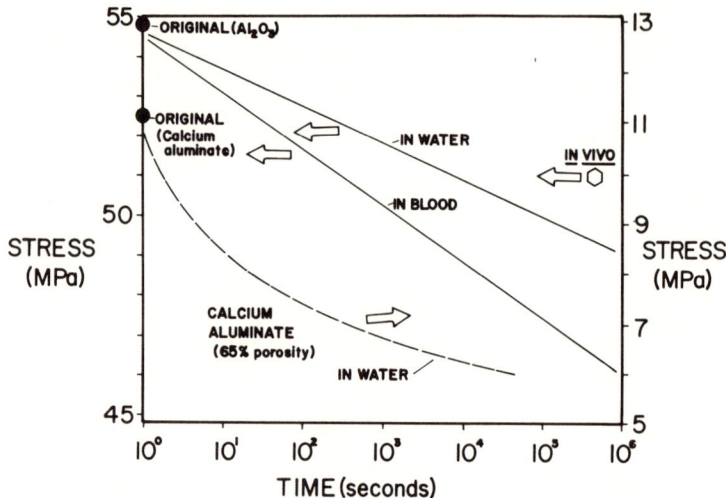

Figure 5-7. Effect of blood and water on the strength of ceramics under static loading conditions. The sample, in bar shape, was loaded by a predetermined stress and the time recorded until fracture. (Redrawn from J. J. Frakes, *Delayed Failure and Ageing of Alumina in Water and Biological Media,* M.S. thesis, University of Illinois, Urbana, Ill., June 1972, and G. N. Schnittgrund, *Static Fatigue of Calcium Aluminate,* M.S. thesis, University of Illinois, Urbana, Ill., September 1971.)

A substantial decrease in strength was also observed when the samples were implanted in animals without any loads. However, the dynamic nature of the *in vivo* condition (the sample moves with the movement of the host) adds one more difficulty to interpreting the result.

Example 5-1

A ceramic (Al_2O_3) is used to fabricate a hip joint. Assume a simple ball-and-socket configuration with a surface contact area of 1.0 cm² and continuous static loading in a simulated condition similar to Figure 5-7. (Extrapolate the data if necessary.)

a. How long will it last if the loading is 70 kg (mass) in water? In blood?
b. Will the implant last longer or shorter with dynamic loading? Give reasons.

Answers

a. Force = mass × acceleration = 70 kg × 9.8 m/s² = 686 N

$$\text{Stress } (\sigma) = \frac{\text{force}}{\text{area}} = \frac{686 \text{ N}}{1 \text{ cm}^2} = \underline{6.86 \text{ MPa}}$$

CERAMIC MATERIALS

From Figure 5-7, $\sigma = 54.7 - 0.95 \log t$. Therefore, $6.86 = 54.7 - 0.95 \log t$

$$t = 10^{50.35} = 2.2 \times 10^{50} \text{ s} = \underline{7 \times 10^{42} \text{ years}} \text{ (forever)}$$

in blood,
$$\sigma = 54.7 - 1.45 \log t$$

Therefore, $t = 10^{37.7} \text{ s} = \underline{1.59 \times 10^{30} \text{ years}}$ (forever).

b. Because the dynamic loading increases the force about 5 to 8 times (Table 12-1), let us assume 8 times, which increases the stress 8-fold. $6.86 \times 8 = \underline{54.88 \text{ MPa}}$; this exceeds the strength of the alumina and failure occurs immediately.

5.4. CARBONS

Although carbons are made of one element, carbon (C), there are many different carbon structures with different properties. The lead of a pencil is graphite, which is one of the carbon structures. Graphite is composed of aggregates of crystallites, which in turn are made of parallel layers of hexagonal lattice structures of elemental carbon, as shown in Figure 5-8. The parallel layers of the crystallites have cross-links between them. The degree of cross-linking determines whether the material is a

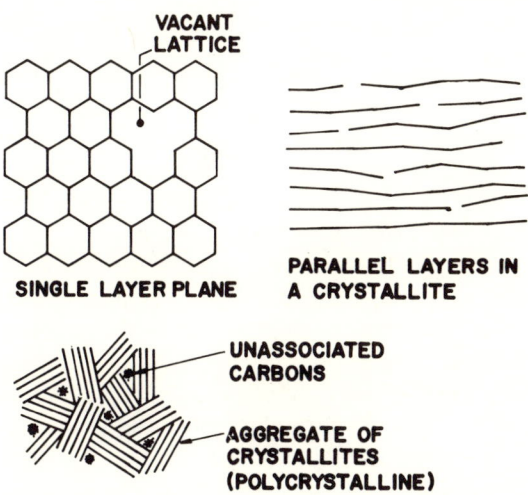

Figure 5-8. Schematic diagrams of the carbon structures. (Redrawn from J. C. Bokros, L. D. La Grange, and G. J. Schoen, *Chemistry and Physics of Carbon,* vol. 9, ed. P. L. Walker, pp. 103–171, Marcel Dekker, New York, 1972, by permission from the publisher.)

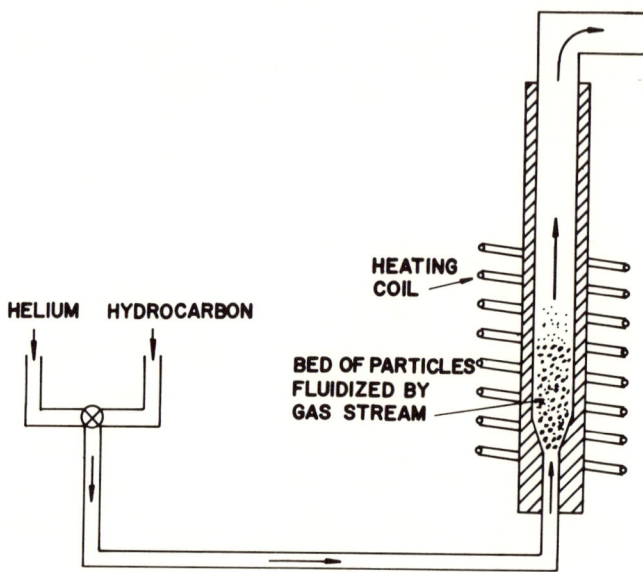

Figure 5-9. Schematic diagram of coating with carbon in a fluidized bed. (Redrawn from J. C. Bokros, *Chemistry and Physics of Carbon,* vol. 5, ed. P. L. Walker, pp. 70–81, Marcel Dekker, New York, 1969, by permission from the publisher.)

lubricating graphite (minimal cross-linking with a large amount of unassociated carbons) or high-strength pyrolytic carbons (maximum cross-linking). Other factors influencing the properties are the size and preferred orientation of the crystallites, microstructure, density, presence of other elements such as silica, and amount of unassociated carbons.

The pyrolytic carbon can be deposited onto a substrate (usually graphite) in a fluidized bed, as shown in Figure 5-9. The hydrocarbons are methane, ethane, propane, etc. and deposited at about 1200°C, which results in a density of about 2.0 g/cm^3. This is the low-temperature isotropic (LTI) pyrolytic carbon used commercially for implants.

The glassy or vitreous carbons are made by pyrolysis of thermosetting polymers such as phenol formaldehyde. The structure lacks a well-defined crystal structure, although a very small region on the order of 50 Å can contain a layered structure of graphite.

The physical properties of various carbon materials are compared in Table 5-4. The modulus of elasticity and compressive strength are substantially lower than they are in ceramic materials. This is due to the large number of flaws introduced during the manufacture of glassy and pyrolytic

Table 5-4. Physical Properties of Carbons[a]

	Carbons		
Properties	Graphite	Glassy	Pyrolytic
Density (g/cm^3)	1.5–1.9	1.5	1.5–2.0
Modulus of elasticity (GPa)	24	24	28
Compressive strength (MPa)	138	172	517

[a] Taken from J. C. Bokros, L. D. LaGrange, and G. J. Schoen, *Chemistry and Physics of Carbon*, vol. 9, ed. P. L. Walker, p. 123, Marcel Dekker, New York, 1972, by permission from the publisher.

carbon. The flaws, in turn, act as stress risers, reducing the strength considerably compared to the theoretical value.

Example 5-2

Calculate the density of a single crystal diamond which has a cubic structure with a bond length of 1.545 Å.

Answer

Since the interatomic distance is one-fourth of the body diagonal, the direction along which the carbon atoms are in contact is

$$3(a/4)^2 = (2r)^2$$
$$a = 4(1.545 \times 10^{-8} \text{ cm})/\sqrt{3}$$
$$= 3.57 \times 10^{-8} \text{ cm}$$

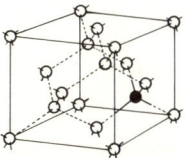

Diamond structure.

There are 4 atoms totally within the unit cell, 3 atoms ($^6/_2$) on the 6 faces, and 1 atom ($^8/_8$) on the 8 corners, or a total 8 atoms per unit cell. Therefore,

$$\text{Density} = \frac{\text{mass/unit cell}}{\text{volume/unit cell}}$$
$$= \frac{(8 \text{ atoms/unit cell})(12 \text{ g/mol})/(6.02 \times 10^{23} \text{ atoms/mol})}{(3.57 \times 10^{-8} \text{ cm})^3/\text{unit cell}}$$
$$= 3.51 \text{ g/cm}^3$$

This value is much higher than the density of graphite, although both are made from the same element, carbon.

PROBLEMS

5-1. Some ceramic materials are used for implants. Explain why they usually are employed for compressive load-carrying devices such as knee and hip joints rather than for tensile loads. Can you use carbons for the same devices?

Answer
The difficulty of predicting the failure load caused by the brittleness of ceramics forced their use as compressive load carriers. Ceramic's tensile strength is also lower than its compressive strength because its micropores act as stress risers.

Carbons can be used in the same devices as ceramics but their strength is lower, although fabrication is easier than with ceramics. The black color of carbons is often considered nonaesthetic.

5-2. Glass ceramics are proposed to have a direct bonding between implant and bone. Propose a model for this direct bonding. How can you prove this?

Answer
It is proposed that the insufficient or excess surface ion concentrations produce negative osteogenesis and that fixation (bonding) results. A summary of the surface chemistry on the bioglass implant with hard tissues is shown below. A thin (500 Å) zone of co-crystallization within the inorganic gel and the collagen produces the bonding [A. E. Clark, L. L. Hench, and H. A. Paschall, *J. Biomed. Mater. Res. 10,* 161 (1976)]. See Figure 8-7 for an electron micrograph of the interface.

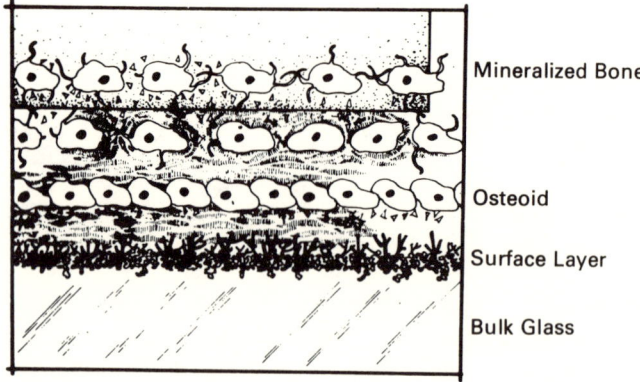

5-3. Some piezoelectric ceramics are being considered as substitutes for hard tissues. It has been shown that the tissues are piezoelectric and that healing can be enhanced by electrical potential.

a. What considerations should be given to the design of a piezoelectric ceramic implant to be used for a bone gap?
b. What type of fixation method can be used?
c. What advantages and disadvantages do metallic implants have as compared with screws, plates, and intramedullary rods?

CERAMIC MATERIALS

Answers
a. If the piezoelectricity of a material can be beneficial in increasing tissue healing, as in the case of electrical stimulation, then (1) the tissue compatibility, (2) the amount of piezoelectric potentials for given loads, and (3) the long-term effects of such stimulation should be studied.
b. The bone gap implant surface can be made porous to allow tissue ingrowth. Immobilization of the implant in the early stage will require screws and rods.
c. Piezoelectric implants can be thought of as *artificial bone*, with properties similar to those of natural bone. Artificial bone will not induce immunogenic reaction, which is the major problem of transplantation.

5-4. Give as many examples as possible of the use of ceramics, including carbons, as implants.

Answer
Acetabular cup and femoral head for hip joint, elbow joint, shoulder joint, toe joint teeth, transcutaneous implant.

5-5. Replot the following figure according to Figure 5-6 and answer the following.

a. Does the polyethylene follow the same curve as that in Figure 5-6?
b. Explain why the porous materials, regardless of their intrinsic properties, follow the same curve of relative strength versus pore volume fraction.

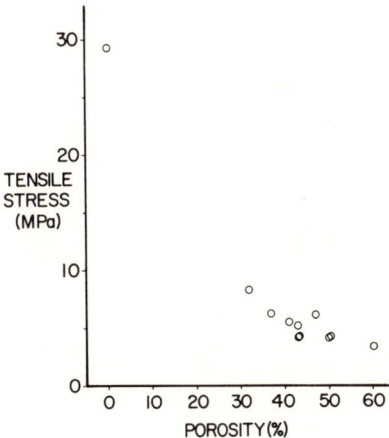

Tensile stress versus porosity of high-density polyethylene.

Answers
a. Yes.
b. The failure rates of porous materials may be the same regardless of the material because the overriding factor in failure is the stress concentration caused by the pores.

FURTHER READING

J. J. Gilman, "The Nature of Ceramics," in *Materials,* eds. D. Flanagan et al., W. H. Freeman & Co., San Francisco, 1967.

W. D. Kingery, H. K. Bowen, and D. R. Uhlmann, *Introduction to Ceramics,* 2nd ed., J. Wiley and Sons, New York, 1976.

W. G. Moffatt, G. W. Pearsall, and J. Wulff, *The Structure and Properties of Materials,* vol. I, *Structure,* chapter 3, J. Wiley and Sons, New York, 1964.

F. Norton, *Elements of Ceramics,* 2nd ed., Addison-Wesley, Reading, Mass., 1974.

L. H. Van Vlack, *A Textbook of Materials Technology,* chapter 10, Addison-Wesley, Reading, Mass., 1973.

L. H. Van Vlack, *Elements of Materials Science,* 2nd ed., chapter 8, Addison-Wesley, Reading, Mass., 1964.

CHAPTER 6

POLYMERIC MATERIALS

Polymers (*poly* = many, *mer* = unit) are linked together by the primary covalent bonding in the main chain backbone with C, N, O, Si, etc., atoms. The simplest example (but not a simple material) is polyethylene, which is derived from ethylene (CH_2=CH_2), in which the carbon atoms share electrons with two other hydrogen and carbon atoms: —$CH_2(CH_2$—$CH_2)_n CH_2$—, where n indicates the number of repeat units.

To make a strong solid the repeat unit n should be well over 10,000, making the molecular weight (M.W.) of the polymer over $1/4$ million grams per mole. This is why the polymers are made of giant molecules. At low molecular weight the material behaves like a wax (paraffin wax used for household candles) and at still lower M.W. as an oil and gas.

The main backbone chain can be of entirely different atoms; for example, polydimethyl siloxane (silicone rubber), —$Si(CH_3)_2[O$—$Si(CH_3)_2]_n O$—. If we substitute the hydrogen atoms of polyethylene with fluorine (F), the resulting material is well known—Teflon® (polytetrafluoroethylene).

Implants made of polymers have several advantages and disadvantages over metals and ceramics. Among their advantages:

1. They can easily be fabricated into many forms of final usage, such as oils, fabrics, films, and/or solids.
2. They are noncorrosive in the body (this does not mean nondegradable) compared to the metals.
3. They bear a close semblance to such natural tissues as collagen, which makes it possible to incorporate other substances by direct bonding, e.g., heparin coating on the surface of polymers for preventing blood clotting.
4. Adhesive polymers can be used as a nonsuturing method of closing wounds or torn organs or luting orthopedic implants in place.

5. The density of polymers is closer to the density of the natural tissues (density of 1 g/cm^3).

The disadvantages are:

1. Low modulus of elasticity and viscoelastic characteristics make the polymers difficult to use for large load-bearing applications.
2. The nature of polymerization makes them biodegradable in the body (they cannot be 100% polymerized).
3. It is very hard to obtain pure medical-grade polymers without such additives as antioxident, antidiscoloring agents, and plasticizers because most polymers are produced in large quantities for other purposes except in very limited cases.

6.1. POLYMERIZATION

To link the molecules (monomers) the monomer must be forced to lose its electrons by the processes of condensation and addition. By controlling the reaction temperature, pressure, and time in the presence of catalyst(s), the degree to which monomers are put together into chains can be manipulated.

6.1.1. Condensation Polymerization

During condensation polymerization, a small molecule such as water will be condensed out of the chemical reaction:

$$R\text{—}NH_2 + R'COOH \rightarrow R'CONHR + H_2O \qquad (6\text{-}1)$$
$$\text{(amine)} \quad \text{(carboxylic acid)} \quad \text{(amide)}$$

This particular process is used to make polyamide (nylon), the first commercial polymer, made initially in the 1930s.

Most natural polymers like cellulose (polysaccharides) and proteins are made by condensation polymerization. Cellulose can be polymerized from the common monosaccharide, glucose, by condensing a water molecule:

(6-2)

Glucose → Cellulose (polysaccharide)

POLYMERIC MATERIALS

Hyarulonic acid, chondroitin, and chondroitin sulfate are important polysaccharides present in connective tissues. These polysaccharides lubricate the joints and fibrous connective tissue layers like collagen and elastin.

Collagen and elastin are proteins composed of amino acids, which are the same as monomers. There are about 30 naturally occurring amino acids which are polymerized into (poly) peptides by the condensation process:

$$2\ H_2N-\underset{R}{\underset{|}{C}}\underset{H}{\overset{H}{|}}-\overset{O}{\overset{\|}{C}}-OH$$

$$\rightarrow H_2N-\underset{R}{\underset{|}{\overset{H}{\overset{|}{C}}}}-\overset{O}{\overset{\|}{C}}-\underset{H}{\underset{|}{N}}-\underset{R}{\underset{|}{\overset{H}{\overset{|}{C}}}}-\overset{O}{\overset{\|}{C}}-OH \xrightarrow{\text{enzyme}} -\underset{R}{\underset{|}{\overset{H}{\overset{|}{C}}}}-\overset{O}{\overset{\|}{C}}-\underset{H}{\underset{|}{N}}-\underset{R'}{\underset{|}{\overset{H}{\overset{|}{C}}}}-\overset{O}{\overset{\|}{C}}-\underset{H}{\underset{|}{N}}-\underset{R''}{\underset{|}{\overset{H}{\overset{|}{C}}}}- \quad (6\text{-}3)$$

Peptide linkage

Some typical condensation polymers and their chemical reactions are given in Table 6-1. One major drawback of condensation polymerization is the tendency for the reaction to cease before the chains grow to a sufficient length. This is because mobility of the chains decreases as polymerization progresses, which results in many short chains. Nylon chains are polymerized sufficiently long enough before mobility decreases and the physical properties of the polymer are preserved. If longer chains are desired, as in the case of vinyl polymers, addition polymerization is utilized.

6.1.2. Addition or Free Radical Polymerization

Addition polymerization can be achieved by rearranging the bond within each monomer. Because each "mer" must share at least two covalent electrons with other mers, the monomer has to have at least one double bond. For example, in ethylene:

$$\underset{H}{\overset{H}{|}}C=\underset{H}{\overset{H}{|}}C \rightarrow -\underset{H}{\overset{H}{|}}\overset{|}{C}-\left(\underset{H}{\overset{H}{|}}\overset{|}{C}-\underset{H}{\overset{H}{|}}\overset{|}{C}\right)_n-\underset{H}{\overset{H}{|}}\overset{|}{C}- \quad (6\text{-}4)$$

The breaking of a double bond can be made with an *initiator*. This is usually a free radical such as benzoyl peroxide:

$$C_6H_5COO-OOC_6H_5 \rightarrow 2C_6H_5COO\cdot \rightarrow 2C_6H_5\cdot + 2CO_2 \quad (6\text{-}5)$$
$$(R\cdot)$$

Table 6-1. Typical Condensation Polymers[a]

Type and interunit linkage	Reaction examples
Polyester $$-\overset{O}{\underset{\|}{C}}-O-$$	$HO(CH_2)_nCOOH \rightarrow HO[-(CH_2)_nCOO-]_mH + H_2O$ $HO(CH_2)_nOH + HOOC(CH_2)_{n'}COOH \rightarrow HO[-(CH_2)_nO\overset{O}{\underset{\|}{C}}(CH_2)_{n'}-\overset{O}{\underset{\|}{C}}O-]_mH + H_2O$ $\begin{array}{c}CH_2OH\\ \| \\ CHOH + HOOC(CH_2)_nCOOH \rightarrow \text{3-dim. network} + H_2O\\ \| \\ CH_2OH\end{array}$
Polyamide $$-\overset{O}{\underset{\|}{C}}-NH-$$	$NH_2(CH_2)_nCOOH \rightarrow H[-NH(CH_2)_nCO-]_mOH + H_2O$ $NH_2(CH_2)_nNH_2 + HOOC(CH_2)_{n'}COOH \rightarrow$ $H[-NH(CH_2)_nNHCO(CH_2)_{n'}CO-]_mOH + H_2O$
Polyurethane $$-O-\overset{O}{\underset{\|}{C}}-NH-$$	$HO(CH_2)_nOH + OCN(CH_2)_{n'}CNO \rightarrow$ $[-O(CH_2)_nOCONH(CH_2)_{n'}NHCO-]_m$
Polyurea $$-NH-\overset{O}{\underset{\|}{C}}-NH-$$	$NH_2(CH_2)_nNH_2 + OCN(CH_2)_{n'}CNO \rightarrow$ $[-NH(CH_2)_nNHCONH(CH_2)_{n'}NHCO-]_m$
Silk fibroin $$-\overset{O}{\underset{\|}{C}}-NH-$$	$NH_2CH_2COOH + NH_2CHROOH \rightarrow$ $H[-NHCH_2CONHCHRCO-]_mOH + H_2O$
Polysiloxane $$\begin{array}{c}R\quad\ \ R\\ \| \quad\ \ \|\\ -Si-O-Si-\\ \| \quad\ \ \|\\ R\quad\ \ R\end{array}$$ $$\begin{array}{c}R\quad\ \ R\\ \| \quad\ \ \|\\ -Si-O-Si-\\ \| \quad\ \ \|\\ R\quad\ \ O\\ \quad\ \ \|\\ R-Si-R\end{array}$$	$\begin{array}{c}CH_3\\ \|\\ HO-Si-OH \rightarrow HO-\left[\begin{array}{c}CH_3\\ \|\\ Si-O\\ \|\\ CH_3\end{array}\right]_m + H_2O\\ \|\\ CH_3\end{array}$ $\begin{array}{cc}CH_3 & CH_3\\ \| & \|\\ HO-Si-OH + HO-Si-OH \rightarrow \text{3-dim. network}\\ \| & \|\\ CH_3 & CH_3\end{array}$

[a] Adapted from F. W. Billmeyer, *Textbook of Polymer Science*, p. 240, Interscience Publishers, New York, 1962.

The initiation can be activated by heat, ultraviolet light, and other chemicals.

The free radicals (initiators) can react with monomers:

$$R\cdot + CH_2=CHX \rightarrow RCH_2-\underset{X}{\overset{H}{\underset{|}{\overset{|}{C}}}}\cdot \quad (6\text{-}6)$$

POLYMERIC MATERIALS

and this free radical can react with another monomer:

$$RCH_2 - \underset{X}{\overset{H}{\underset{|}{C}}} \cdot + CH_2 = CHX \rightarrow RCH_2 - CHX - CH_2 - \underset{X}{\overset{H}{\underset{|}{C}}} \cdot \qquad (6\text{-}7)$$

and the process can continue indefinitely.

This process is called a propagation and can be written in a short form:

$$\begin{aligned} R \cdot + M &\rightarrow RM \cdot \\ RM \cdot + M &\rightarrow RMM \cdot \end{aligned} \qquad (6\text{-}8)$$

where M represents any monomer.

The propagation process can be terminated by combining two free radicals, transfer, and disproportionate processes:

$$RM_nM \cdot + R \cdot (\text{or } RM \cdot) \rightarrow RM_{n+1}R \text{ (or } RM_{n+2}R) \qquad (6\text{-}9)$$
$$RM_nM \cdot + RH \rightarrow RM_{n+1}H + R \cdot \qquad (6\text{-}10)$$
$$RM_nM \cdot + \cdot MM_nR \rightarrow RM_{n+1} + M_{n+1}R \qquad (6\text{-}11)$$

$$\left(\begin{array}{c} \overset{H}{\underset{X}{\underset{|}{C}H_2C}} \cdot + \cdot \overset{H}{\underset{X}{\underset{|}{C}}}-CH_2- \rightarrow -CH_2\overset{H}{\underset{X}{\underset{|}{C}H}} + \overset{H}{\underset{X}{\underset{|}{C}}}=CH- \end{array} \right)$$

Some of the commercially important monomers for addition polymers are given in Table 6-2.

The degree of polymerization (DP) is one of the most important

Table 6-2. Monomers for Addition Polymerization

Monomer	Addition	Monomer	Addition
Vinyl chloride	$(CH_2{=}CHCl)$	Vinyl acetate	$(CH_3COOCH{=}CH_2)$
Styrene	$(CH_2{=}CH{-}C_6H_5)$	Vinylidene chloride	$(CH_2{=}CCl_2)$
Methylacrylate	$(CH_2{=}CH{-}COOCH_3)$	Methylmethacrylate	$\left(CH_2{=}C\diagup^{CH_3}_{\diagdown COOCH_3} \right)$
Acrylonitrile	$(CH_2{=}CH{-}CN)$		

values determining the physical properties. It is defined as the number of mers per molecule. Each molecule may have small or large numbers of mers, depending on the condition of the polymerization. Therefore, we speak of average degree of polymerization on average molecular weight, $\overline{M}$. The relationship between molecular weight and DP can be expressed as

$$\overline{M} = \text{DP} \times \text{M.W. of mer} \qquad (6\text{-}12)$$

The number average molecular weight can be calculated according to the number of molecules (Xi) in each molecular weight fraction (MWi),

$$\overline{Mn} = \frac{\Sigma(Xi \cdot MWi)}{\Sigma Xi} \qquad (6\text{-}13)$$

Similarly, the weight average molecular weight can be calculated according to the weight fraction (Wi) in each molecular weight fraction,

$$\overline{Mw} = \frac{\Sigma(Wi \cdot MWi)}{\Sigma Wi} \qquad (6\text{-}14)$$

The ratio of the weight average and the number average molecular weight ($\overline{Mw}/\overline{Mn}$) indicates the uniformity of the molecule size distribution. This is why $\overline{Mw}/\overline{Mn}$ is called polydispersity. If the ratio is 1, then only one size molecule exists throughout the polymer. Usually it varies between 1.5 and 2.5. Because uniform molecular size distribution is an important factor for physical properties, one should try to obtain a polymer with a smaller polydispersity.

Example 6-1

The following data were obtained for a polypropylene:

Mean M.W. (g/mol)	Weight (g)	Number fraction
50,000	1.0	0.1
40,000	2.0	0.5
20,000	1.0	0.4

a. What are the $\overline{Mn}$ and $\overline{Mw}$?
b. How many molecules are there in 1 gram?
c. What is the degree of polymerization?

Answers

a.
$$\overline{M_n} = \frac{50{,}000 \times 0.1 + 40{,}000 \times 0.5 + 20{,}000 \times 0.4}{0.1 + 0.5 + 0.4} = \underline{33{,}000 \text{ g/mol}}$$
$$\overline{M_w} = \frac{50{,}000 \times 1.0 + 40{,}000 \times 2.0 + 20{,}000 \times 1.0}{1.0 + 2.0 + 1.0} = \underline{37{,}500 \text{ g/mol}}$$

b.
$$\frac{\text{Number of molecules}}{\text{gram}} = \frac{6.02 \times 10^{23} \text{ molecules/mol}}{37{,}500 \text{ g/mol}} = \underline{1.6 \times 10^{19}}$$

c. The molecular weight of propylene monomer is 42 g/mol; therefore,

$$\text{DP} = \frac{37{,}500 \text{ g/mol}}{42 \text{ g/mol}}$$
$$= \underline{893}$$

6.1.3. Solid State of Polymers

As the molecular chains become longer by polymerization, their relative mobility decreases. The chain mobility is also related to mechanical strength, which in turn is directly proportional to the molecular weight, as shown in Figure 6-1.

Such linear polymers as polyethylene and polyamide (nylon) cannot be crystallized as easily as metals because the individual chains are long, making the relative motion difficult. Another factor is the length of chains, which varies from chain to chain, creating many defects because the chain ends act as defects (similar to the edge dislocation).

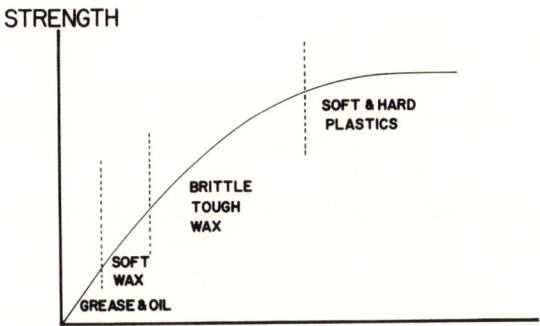

Figure 6-1. Strength versus molecular weight of polyolefins (mainly polyethylene and propylene). The strength increase is mainly caused by the decrease in the relative motion of chains as they become longer with higher molecular weight.

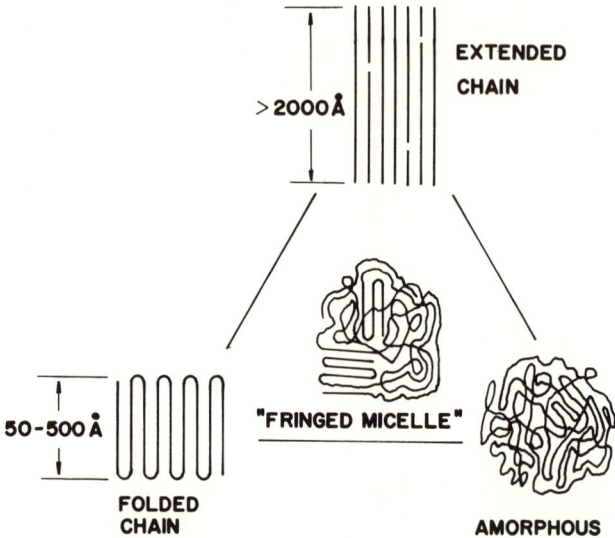

Figure 6-2. Two-dimensional representation of polymer structure. Generally, the fibers have more extended chains, single crystals have folded chains, and glassy polymers have amorphous structures. The semicrystalline polymers are thought to have the "fringed micelle" structure. (Modified from B. Wunderlich, *Crystals of Linear Macromolecules,* ACS Audio Course, American Chemical Society, Washington, D.C., 1973.)

Nevertheless, linear polymers do crystallize from melt or solution. Generally, a complete crystallization is impossible and the resulting structure is a mixture of crystalline and noncrystalline regions, as shown in Figure 6-2. The arrangement of chains in the crystalline region is believed to be a combination of folded and extended chains in the amorphous matrix. The chain folds are necessary to explain a single crystal structure in which the thickness is too small to accommodate the length of the chain (X-ray and electron diffraction of the crystals shown in the chains should be aligned in the direction of chain length).

The extended chain configuration can occur by alignment of chains, which is the lowest energy state for the solid state. The orientation of chains can be accomplished easily by drawing through fine holes from polymer melt or solution. Even this process does not prevent chain folds entirely. The ends of folds, as in chain ends, act as defects, thus lowering strength. To some degree the chain folds can be prevented by stretching at high temperature (below Tm), which results in higher-strength fibers.

The three-dimensional network polymers, such as (poly)phenolformaldehyde (Bakelite®), do not usually crystallize. This type of structure can be considered as if chains of amorphous polymers are cross-linked

POLYMERIC MATERIALS

randomly. In fact, the solid isoprene rubber is made by cross-linking with sulfur (vulcanization). Too much cross-linking may restrict the chain movement severely, making the rubber very hard. Because of this lack of freedom of chains, the three-dimensional network polymers do not melt; instead, they burn.

Example 6-2

The unit cell structure of polyethylene is orthorhombic, as shown. Calculate the density for a 100% crystalline polyethylene.

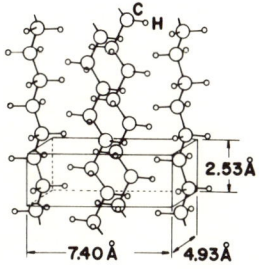

Unit cell structure of a polyethylene.

Answer

One unit cell contains 2 mers and the molecular weight of each mer is 28 g/mol. Therefore,

$$\text{Density} = \frac{(2 \text{ mers})(28 \text{ g/mol})/(6.02 \times 10^{23} \text{ mers/mol})}{(7.40 \times 10^{-8} \text{ cm})(4.93 \times 10^{-8} \text{ cm})(2.53 \times 10^{-8} \text{ cm})}$$
$$= \underline{1.01 \text{ g/cm}^3}$$

Example 6-3

The average end-to-end distance ($\overline{L}$) of a chain of an amorphous polymer can be expressed as $\overline{L} = l \sqrt{m}$, where l is the interatomic distance (1.54 Å for C—C) and m is the number of bonds. If the average molecular weight of polystyrene is 20,800 g/mol, what is the average end-to-end distance ($\overline{L}$) of a chain?

Answer

Molecular weight of polystyrene = 104 g/mol

$$\frac{20,800}{104} = 200 \text{ mers} = 400 \text{ bonds}$$

Therefore, $\overline{L} = l \sqrt{m} = 1.54 \sqrt{400} = 30.8$ Å. The average end-to-end distance of a chain for amorphous polymers is quite short, although their total bond distance is quite long (~600 Å in this example). In a crystalline solid the chains are not randomly distributed; thus this calculation does not apply.

6.2. EFFECT OF STRUCTURAL MODIFICATION ON PROPERTIES

The physical properties of polymers can be affected in many ways. By this means polymers can be tailored to meet their end use. Basically, the chemical composition and arrangement of chains will have a great effect on the final properties.

6.2.1. Effect of Molecular Weight and Composition

The molecular weight and its distribution have a great effect on the properties of a polymer since its rigidity is primarily caused by the immobilization of the chains. This is because the chains are like cooked threads of spaghetti in a bowl. By increasing molecular weight the polymer chains become longer and less mobile and a more rigid material results. Equally important is that all chains should be equally long; any short chains will act as plasticizers, which in turn decrease the melting and glass transition temperatures, rigidity (modulus of elasticity), density, etc.

Another obvious way of changing properties is to change the chemical composition of the backbone or side chains. Substituting the backbone carbon of a polyethylene with divalent oxygen or sulfur will decrease melting and glass transition temperatures because the chain becomes more flexible as a result of increased rotational freedom:

(6-15)

But the opposite effect can be achieved by substituting the backbone chains with a rigid molecule like benzene:

(6-16)

Polyethylene terephthalate (polyester, Dacron®)

6.2.2. Effects of Side-Chain Substitution, Cross-Linking, and Branching

Increasing the size of side groups in linear polymers like polyethylene will decrease the melting temperature as a result of the lesser perfection of molecular packing, i.e., decreased crystallinity. This effect is seen until the side group itself becomes large enough to hinder the movement of the main chain as shown in Table 6-3. Very long side groups can be thought of as branches.

Cross-linking of the main chains is in effect similar to side-chain substitution with a small molecule; i.e., it lowers the melting temperature. This is due to the interference of the cross-linking, which can decrease the mobility of the chains, resulting in further retardation of the crystallization rate. In fact, a large degree of cross-linking can completely prevent crystallization.

Example 6-4

Assuming that the density of a supercooled liquid polyethylene is 0.9 g/cm³, what is the percent crystallinity if the densities for low- and high-density polyethylene are (a) 0.92 and (b) 0.97 g/cm³?

Answers

a. $\dfrac{0.92 - 0.9}{1.01 - 0.9} = \dfrac{0.02}{0.11} = \underline{18\% \text{ crystalline}}$

b. $\dfrac{0.97 - 0.9}{1.01 - 0.9} = \dfrac{0.07}{0.11} = \underline{64\% \text{ crystalline}}$

The low-density polyethylene contains side chains and branches.

Table 6-3. Effect of Side-Chain Substitution on Melting Temperature in Polyethylene

Side chain	$Tm(°C)$
—H	140
—CH$_3$	165
—CH$_2$CH$_3$	124
—CH$_2$CH$_2$CH$_3$	75
—CH$_2$CH$_2$CH$_2$CH$_3$	−55
—CH$_2$CH(CH$_3$)—CH$_2$CH$_3$	196
—CH$_2$—C(CH$_3$)$_2$—CH$_2$CH$_3$	350

6.3. PROPERTIES OF POLYMERS

Conventionally, the polymers are classified into two different categories depending on their thermal properties, i.e., thermoplastic and thermosetting polymers. The thermoplastic polymers are linear in molecular structure and can be remelted. Most polymers made by free radical reactions (addition polymerization) are thermoplastic. The thermosetting polymers cannot be remelted because their chains are made in a three-dimensional network. Therefore, the chains cannot move unless they are cut off from each other rather than sliding as in the linear polymers.

Rubbers are linear polymers with cross-links between chains. The cross-linking can be achieved either by introducing a chemical agent such as sulfur or by branching between chains as in the case of low-density polyethylene. Although natural rubbers, poly(cis-)isoprene,

$$-H_2C \underset{\underset{CH_3}{\diagup}}{\diagdown} C=C \underset{\underset{CH_2-}{\diagdown}}{\diagup} H$$

are more familiar, an inorganic silicone rubber is more widely used for implantation. Silicone rubber is made from silicone polymers, which in turn are made by condensation polymerization. One type of widely used silicone polymer for making medical-grade silicone rubber is dimethyl siloxane, which is made from dimethyldichlorosilane:

$$Cl-\underset{\underset{CH_3}{|}}{\overset{\overset{CH_3}{|}}{Si}}-Cl$$

which in turn is made from pure silica (SiO_2) reacting with methylchloride (CH_3Cl). The polymerization can be shown as follows:

$$HO-\underset{\underset{CH_3}{|}}{\overset{\overset{CH_3}{|}}{Si}}-O[H + HO]-\underset{\underset{CH_3}{|}}{\overset{\overset{CH_3}{|}}{Si}}-OH \rightarrow -O\left(\underset{\underset{CH_3}{|}}{\overset{\overset{CH_3}{|}}{Si}}-O\right)_n \underset{\underset{CH_3}{|}}{\overset{\overset{CH_3}{|}}{Si}}-O- + nH_2O \quad (6\text{-}17)$$

The cross-linking or vulcanization can be achieved through heat (heat-vulcanizing type) or chemical agents, which can be accomplished at room temperature (room temperature–vulcanizing types or RTVs) as shown below:

$$\text{HO}-\underset{\underset{CH_3}{|}}{\overset{\overset{CH_3}{|}}{Si}}-O\left(\underset{\underset{CH_3}{|}}{\overset{\overset{CH_3}{|}}{Si}}-O\right)_n\underset{\underset{CH_3}{|}}{\overset{\overset{CH_3}{|}}{Si}}-OH + \underset{\text{Propyl orthosilicate}}{Si(OCH_2-CH_2-CH_3)_4} \xrightarrow{\text{Catalyst}}$$

$$\rightarrow\rightarrow\rightarrow -O-\underset{\underset{CH_3}{|}}{\overset{\overset{CH_3}{|}}{Si}}\underset{\text{\textemdash\textemdash\textemdash\textemdash\textemdash\textemdash\textemdash\textemdash}}{[-OH-CH_3CH_2CH_2-]}-O-\underset{\underset{O}{|}}{\overset{\overset{O}{|}}{Si}}-O\underset{\text{\textemdash\textemdash\textemdash\textemdash\textemdash\textemdash\textemdash\textemdash}}{[-CH_2CH_2CH_3-OH-]}\underset{\underset{CH_3}{|}}{\overset{\overset{CH_3}{|}}{Si}}-O-$$

with additional cross-linking structure including:

$$-O-\underset{\underset{CH_3}{|}}{\overset{\overset{CH_3}{|}}{Si}}-O-$$ \quad (6-18)

where the central chain contains $-O-CH_2-CH_2-CH_3$ groups and $O-H-HO-Si$ linkages with CH_3 substituents.

The cross-linking for the silicone rubber increases the strength and melting temperature, contrary to such other polymers as polyethylene, as discussed previously. This is largely because the rubber structure is essentially an amorphous state (but not supercooled liquid) in which the molecules are linked together via cross-links, resulting in a three-dimensional network structure.

The degree of cross-linking (or cross-link density) is precisely controlled to tailor for the final application. Too high cross-link density may result in too-rigid (nonstretchable) rubber.

6.3.1. Mechanical Properties

As discussed in Chapter 2, polymers are viscoelastic; therefore, the mechanical properties depend greatly on testing conditions. Polymers are especially sensitive to temperature and rate of loading. Generally, the strength of a polymer depends on its molecular structure, i.e., degree of crystallization, orientation, and polymerization. It is also true that by orienting chains in a polymer like nylon, one can achieve a specific strength (strength/density) approaching that of steel in the oriented direction.

Table 6-4. Mechanical Properties of Some Polymers

Polymer	Elastic modulus (MPa)	Tensile stress (MPa)	Fracture elong. (%)	Density (g/cm³)
Polyethylene (high density)	500	20–40	400–500	0.95
Polyethylene (low density)	150	7–17	400–600	0.90
Polypropylene	1500	30–40	50–500	0.904
Polyvinyl chloride (rigid)	3000	40–55	400	1.35–1.45
Polyethylene terephthalate	—	150–250	70–130	1.40
Polyamide (nylon 6)	500	80–150	20–225	1.12–1.14
Natural rubber	2.5–100	10–50	100–700	0.93
Silicone rubber	10	6	50–800	0.98
Polytetrafluoroethylene	500	17–28	320–350	2.15–2.20
Polymethylmethacrylate	3000	70	2.5–5.4	1.18

The higher crystallinity of a linear polymer results from the better packing of the chains, which results in greater strength. In polyethylene the low-density variety has about 50 percent crystallinity and the high-density one has above 70 percent. Consequently, the former has a lower melting point, lower tensile strength and modulus, and a higher elongation to fracture than the latter. Table 6-4 shows some typical values of mechanical properties of various polymers. As can be seen, their strength, elastic modulus, and density are much lower than metals and ceramics.

Example 6-5

Porous polyethylene (100-500 μm diameter, interconnecting pores) is tested in the form of rods with a 3.4-mm diameter *in vitro* and *in vivo* for two months. Typical force-elongation curves are obtained as follows, with a 1-cm gage length.

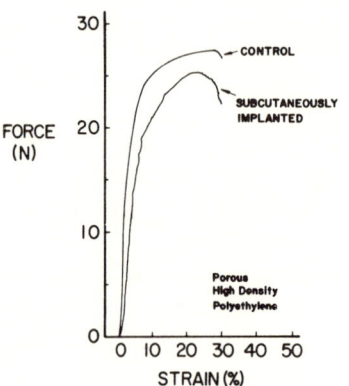

Force versus strain of control and implanted porous high-density polyethylene.

POLYMERIC MATERIALS

a. Calculate the modulus of elasticity and tensile strength for both curves.
b. Explain why the curve for the *in vivo* testing sample is not as smooth as the control sample and has lower strength than the control.

Answers

a. Tensile stresses, area = $\pi(1.7)^2 \times 10^{-6}$ m² = 9.08×10^{-6} m²

$$\text{for control} \quad \sigma_c = \frac{27.5 \text{ N}}{9.08 \times 10^{-6} \text{ m}^2} = \underline{3.03 \text{ MPa}}$$

$$\text{for implanted} \quad \sigma_i = \frac{25.5 \text{ N}}{9.08 \times 10^{-6} \text{ m}^2} = \underline{2.81 \text{ MPa}}$$

Modulus of elasticity

$$\text{for control} \quad E_c = \frac{30 \text{ N}}{9.08 \times 10^{-6} \text{ m}^2 \cdot 0.02} = \underline{165 \text{ MPa}}$$

$$\text{for implanted} \quad E_i = \frac{15 \text{ N}}{9.08 \times 10^{-6} \text{ m}^2 \cdot 0.03} = \underline{55 \text{ MPa}}$$

b. The implanted sample has ingrown tissues which are hard to remove from the porous surface; when tested they exhibit a random tearing that results in the uneven curve. The porous polymer is deteriorated by implantation, resulting in lower strength than the control.

6.3.2. Thermal Properties

Everyone is familiar with the fact that most plastics cannot withstand high temperatures. This is because the rigidity of polymers is directly proportional to the mobility of the chains, as mentioned previously. Most linear polymers melt at about 150–300°C. The glassy polymers such as PMMA (polymethylmethacrylate) and polycarbonate (bisphenol A) have glass-transition temperatures (or softening temperatures) of 60–150°C. The glassy polymers are formed by freezing the chains before they are crystallized, as shown in Figure 3-12.

The amount of supercooling from the liquid melt depends on the rate of cooling—i.e., the slower the rate, the more supercooling, resulting in a lower Tg. The glass transition temperature is the temperature where the material freezes and becomes solid like glass. Semicrystalline polymers like polyethylene have both Tg and Tm. Although the Tg of polyethylene is well below room temperature, it behaves like a solid because the crystalline portion is holding the whole chains together. The glass transition temperatures of semicrystalline polymers range from one-half to two-thirds of the melting temperature in absolute temperature units (K).

6.4. DETERIORATION OF POLYMERS

The deterioration of polymers usually results from several factors, which can be divided into chemical, thermal, and physical categories. These factors may act synergistically, accelerating the deterioration process. Usually the deterioration affects the main chain, the side groups, and their original molecular arrangement.

6.4.1. Chemical Effects

If a linear polymer is undergoing deterioration the main chain will usually be randomly scissioned. Sometimes depolymerization occurs that differs from the random chain scission because this process is the inverse of the disproportionate chain termination of addition polymerization [see equation (6-11)].

Cross-linking of a linear polymer may result in deterioration. An example of this is the low-density polyethylene. On the other hand, if cross-linking is broken by the oxygen or ozone attack on (poly)isoprene rubber, the rubber becomes brittle and cracks develop.

It is also undesirable to change the nature of bonds as in the case of polyvinyl chloride and polyvinyl acetate.

$$\begin{array}{c}\text{H H H H H} \\ | \ | \ | \ | \ | \\ -\text{C}-\text{C}-\text{C}-\text{C}-\text{C}- \\ | \ | \ | \ | \ | \\ \text{H Cl H Cl H} \\ \text{Polyvinyl chloride}\end{array} \rightarrow \begin{array}{c}\text{H H} \quad \text{H H} \\ | \ | \quad\ | \ | \\ -\text{C}-\text{C}-\text{C}=\text{C}-\text{C}- + \text{HCl} \\ | \ | \quad\ | \\ \text{H H Cl} \quad \text{H}\end{array} \quad (6\text{-}19)$$

$$\begin{array}{c}\text{H H} \\ | \ | \\ -\text{C}-\text{C}- \\ | \ | \\ \text{H OCOCH}_3 \\ \text{Polyvinyl acetate}\end{array} + \text{H}_2\text{O} \rightarrow \begin{array}{c}\text{H H} \\ | \ | \\ -\text{C}-\text{C}- \\ | \ | \\ \text{H OH} \\ \text{Polyvinyl alcohol (water soluble)}\end{array} + \text{CH}_3\text{COOH} \quad (6\text{-}20)$$

The by-products of the degradation can be very irritable to tissues because in this case they are acids.

6.4.2. Thermal Effects during Sterilization

In conjunction with sterilization the thermal effect plays an important role in polymer deterioration. In dry sterilization the temperature varies from 160 to 190°C. This range is above the melting or softening temperature of many linear polymers, such as polyethylene and polymethylmethacrylate. In polyamide (nylon), oxidation will occur at the dry sterilization temperature, although it is below its melting temperature. The only polymers that can safely be dry-sterilized are polytetrafluoroethylene (Teflon®) and silicone rubber.

Steam sterilization (autoclave) is performed under high steam pressure at relatively low temperatures (120–135°C). However, if the polymer is subject to attack by water vapor, this method cannot be employed. Polyvinyl chloride, polyacetals, polyethylenes (low-density variety), and polyamides (nylons) belong to this category.

Chemical agents, such as ethylene and propylene oxide gases, and phenolic and hypochloride solutions, are widely used for sterilizing polymers because they can be used at low temperatures. Chemical sterilization takes more time and costs more. Sometimes chemical agents cause polymer deterioration even though sterilization is done at room temperature. However, the time of exposure is relatively short (overnight) and most polymeric implants can be easily sterilized with this method.

Radiation sterilization using the isotope cobalt 60 can also deteriorate polymers because at high dosage the polymer chains can be broken and recombined. At high dosage (above 10^6 Gy) polyethylene becomes a brittle, hard material because of a combination of random chain scission and cross-linking.

6.4.3. Mechanochemical Effect

It is well known that cyclic or constant loading deteriorates polymers. This effect can be accelerated if the polymer is subjected simultaneously to chemical and mechanical activation processes. Thus if the polymer is stored in water or saline solution, its strength will decrease as shown in Figure 6-3. Another reason for the decrease is the plasticizing effect of the water molecules at higher temperature. However, the plasticizing effect compensates for the deleterious effect of saline solution when loaded cyclically, resulting in no differences between the samples stored in saline or in air (Fig. 6-4).

6.4.4. Deterioration of Polymers *in Vivo*

Even though the material implanted inside the body is not subjected to light, radiation, oxygen, ozone, and temperature variations, the body environment is very hostile and all polymers start to deteriorate as soon as they are implanted. The most probable cause of deterioration is ionic attack (especially by the hydroxyl ion, OH^-) and dissolved oxygen. Enzymatic degradation may also play a significant role if the implant is made from natural polymeric materials such as reconstituted collagen.

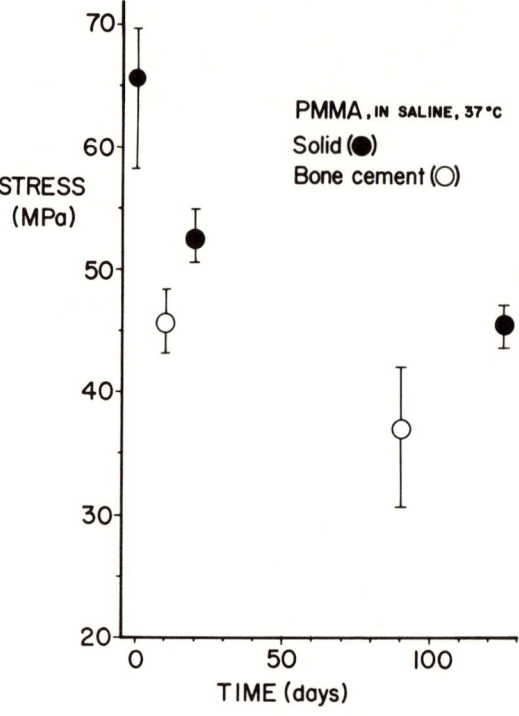

Figure 6-3. Stress versus time for polymethylmethacrylate in saline solution at 37°C. Note the large decrease in tensile strengths for both solid and porous bone cement. (Unpublished data of T. Parchinski, G. Cipolletti, and F. W. Cooke, Clemson University, 1977.)

It is safe to predict that if a polymer deteriorates in physiological solution *in vitro,* it will also deteriorate *in vivo.* Most hydrophilic (water-loving) polymers such as polyamides and polyvinyl alcohol will react with body water and undergo rapid deterioration. The hydrophobic (water-hating) polymers such as polytetrafluoroethylene (Teflon®) and polypropylene are less prone to deteriorate *in vivo.*

The deterioration products may also induce tissue reaction as demonstrated with polyamides (nylons). As in *in vitro* deterioration, the original physical properties will be changed if the implant deteriorates. For example, polyolefins (polyethylene and polypropylene) will lose their flexibility and become brittle. For polyamides the amorphous region is selectively attacked by water molecules which act as a plasticizer, making them more flexible. Table 6-5 shows the effects of implantation on several polymers.

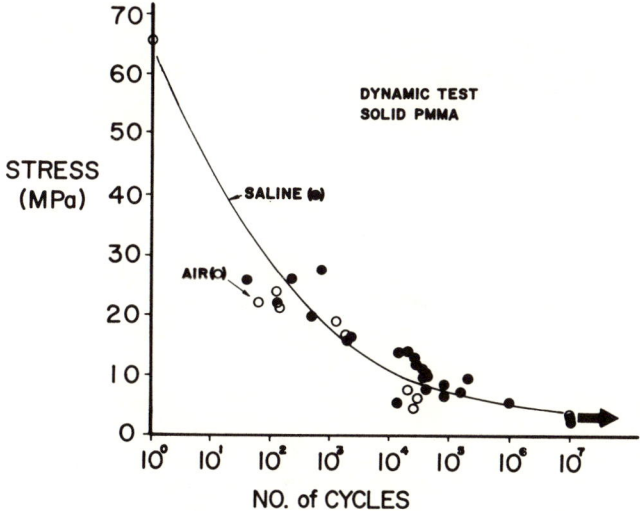

Figure 6-4. Fatigue test of solid polymethylmethacrylate. Note that the *S-N* curve is the same for both untreated samples and samples soaked in saline solution at 37°C. (After T. Parchinski and F. W. Cooke, Clemson University, 1977.)

Table 6-5. Effect of Implantation on Polymers

Polymer	Effect
Polyethylene	Low-density forms: absorb some lipids, lose tensile strength. High-density forms: inert, no deterioration occurs.
Polypropylene	Generally no deterioration.
Polyvinyl chloride (rigid)	Tissue reaction; plasticizers may leach out and become brittle.
Polyethylene terephthalate (polyester)	Susceptible to hydrolysis and loss of tensile strength.
Polyamides (nylons)	Absorb water and irritate tissue; lose tensile strength rapidly.
Silicone rubber	No tissue reaction; very little deterioration.
Polytetrafluoroethylene	Solid specimens are inert. If fragmented, irritation will occur.
Polymethylmethacrylate	Rigid form: crazing, abrasion, and loss of strength by heat sterilization. Cement form: high heat generation during polymerization may damage tissues.

PROBLEMS

6-1. A sample of methylmethacrylate

$$\left(CH_2=C\begin{array}{c} CH_3 \\ \diagup \\ \diagdown \\ COOCH_3 \end{array} \right)$$

is polymerized; the resulting polymer has a DP of 1000. Draw the structure for the repeating unit in the polymer and calculate the polymer molecular weight.

Answer

$$\begin{array}{ccccccc}
 & H & CH_3 & & H & CH_3 & & H & CH_3 \\
 & | & | & & | & | & & | & | \\
-&C&-C-& &C&-C-& &C&-C- \\
 & | & | & & | & | & & | & | \\
 & H & C=O & & H & C=O & & H & C=O \\
 & & | & & & | & & & | \\
 & & O & & & O & & & O \\
 & & | & & & | & & & | \\
 & & CH_3 & & & CH_3 & & & CH_3 \\
\end{array}$$

M.W. of the repeating unit $= 12 \times 5 + 1 \times 8 + 16 \times 2 = 100$ g/mol, $\overline{Mn} = $ DP $\times$ M.W. of the repeating unit $= 1000 \times 100 = \underline{1 \times 10^5 \text{ g/mol}}$

6-2. Which of the following compounds form polymers?

a. $CH_2=CH_2$

b. $HC=CH_2$
     |
     Cl

c. $Cl-C=CH_2$
        |
        Cl

d. $Cl_2C=Cl_2$

e. $H_2C \overset{O}{-\!\!\!\triangle\!\!\!-} CH_2$

f. (benzene ring)

g. $\underset{\underset{\text{(phenyl)}}{|}}{\overset{CH_3}{\underset{|}{C}}}=CH_2$

Answers
a. Polyethylene
b. Polyvinyl chloride
c. Polyvinylidene chloride
e. Polyethylene oxide
g. Methyl-substituted polystyrene

POLYMERIC MATERIALS

6-3. For a Voigt model in which the spring and the dashpot are in parallel, derive the following equations:

a. The differential equation $\eta \, d\epsilon/dt + E = \sigma$

b.
$$\epsilon = (\sigma/E)[1 - \exp(-E/\eta)t] \quad \text{for} \quad t < t_1$$
$$\epsilon = \epsilon_1 \exp[-(E/\eta)t] \quad \text{for} \quad t > t_1$$

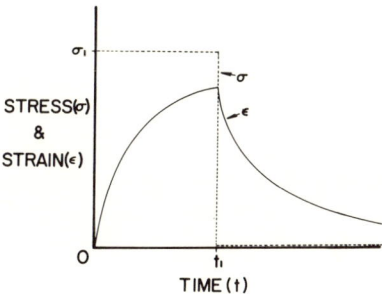

Answers

a. The differential equation is derived in equation (2-19):

$$\underline{\sigma = E\epsilon + \eta \, d\epsilon/dt} \tag{1}$$

b. For $t < t_1$.

Solving for ϵ from the differential equation (1) we have homogeneous solution, i.e.,

$$d\epsilon/dt + (E/\eta)\epsilon = 0$$

$$\int d\epsilon/\epsilon = \int -(E/\eta) \, dt = -(E/\eta) \int dt$$

Therefore, $\ln \epsilon = -(E/\eta)t + C_1$; hence

$$\epsilon = C_2 \exp[-(E/\eta)t] \tag{2}$$

where C_1 and C_2 are constants. Solving for a particular solution by making the stress constant (σ_o),

$$\sigma_o/\eta = d\epsilon/dt + (E/\eta)\epsilon$$

However,

$$\epsilon_{\text{total}} = \epsilon_{\text{dashpot}} = \epsilon_{\text{spring}} = \sigma/E = \sigma_o/E = \text{constant}$$

Therefore, $d\epsilon/dt = 0$;

$$K/\eta = (E/\eta)\epsilon \quad \text{and} \quad \epsilon = \sigma_0/E \tag{3}$$

Combining equations (2) and (3),

$$\epsilon = C_2 \exp[-(E/\eta)t] + \sigma_0/E \tag{4}$$

Since $t = 0$, $\epsilon = 0$; therefore,

$$C_2 = -\sigma_0/E \tag{5}$$

From equations (4) and (5),

$$t < t_1, \quad \epsilon = (\sigma_0/E)\{1 - \exp[-(E/\eta)t]\} \tag{6}$$

At time $t = t_1$, the stress is removed and the elements are stretched to $\epsilon = \epsilon_1$; therefore, from equation (1), $\eta\, d\epsilon/dt + \epsilon = 0$ and the solution is equation (2). Since at this time $\epsilon = \epsilon_1$, the solution can be expressed as

$$t > t_1, \quad \epsilon = \epsilon_1 \exp[-(E/\eta)t] \tag{7}$$

6-4. What are the viscosity and shear modulus for a polymer which behaves as a Voigt model if the shear strains are as follows?

1 h	0.0060
2 h	0.0084
10 h	0.010
20 h	0.010

The shear stress is 10^6 Pa.

Answer
From the differential equation of problem 6-3,

$$\sigma = E\epsilon + \eta\, d\epsilon/dt$$

After 10 h the strain rate ($d\epsilon/dt$) becomes zero; thus $\sigma = E\epsilon$,

$$E = \frac{10^6 \text{ Pa}}{0.01} = \underline{10^8 \text{ Pa}}$$

POLYMERIC MATERIALS

When $\epsilon = 0.006$, the time is 1 h (3600 s)

$$0.006 = \frac{10^6}{10^8} [1 - \exp(-3600/\tau)]$$
$$\tau = 4000 \text{ s}$$

$$\eta = \tau E = 4000 \times 10^8 = \underline{4 \times 10^{11} \text{ Pa} \cdot \text{s}}$$

6-5. An applied strain of 0.4 produces an immediate stress of 10 MPa in a piece of rubber, but after 42 days the stress is only 5 MPa.
a. What is the relaxation time?
b. What is the stress after 90 days?

Answers
Since $\sigma = \sigma_o \exp(-t/\tau)$,

$$\frac{5}{10} = \exp(-42/\tau)$$

a. $$\underline{\tau = 60 \text{ days}}$$

$$\sigma = 10 \exp(-90/60)$$

$$\underline{\sigma = 2.86 \text{ MPa}}$$

6.6. A hip joint is made of a solid polymethylmethacrylate with a contact surface area of 1 cm² (see Figure 6-4).
a. How many cycles will it last if a loading is 70 kg (mass) in saline solution?
b. Compare this simulated situation with the actual implant.

Answers
a. From Figure 6-4, the stress can be approximated by

$$\sigma = 66 - 8.714 \log t \quad \text{and} \quad \sigma = \frac{70 \times 9.8 \text{ N}}{1 \text{ cm}^2} = 6.86 \text{ MPa}$$
$$6.86 = 66 - 8.714 \log t$$

Therefore,
$$t = \underline{6 \times 10^6 \text{ cycles}}$$

b. In humans the mass is divided by two legs, hence the stress is halved. However, the dynamic nature of the loading will make the load about 4–7 times that of the static loading (see Chapter 11) thus making the polymethylmethacrylate unusable as a main load transmitting material.

FURTHER READING

T. Alfrey and E. F. Gurnee, *Organic Polymers,* Prentice-Hall, Englewood Cliffs, N.J., 1967.
F. W. Billmeyer, Jr., *Textbook of Polymer Science,* 2nd edition, J. Wiley and Sons, New York, 1971.
B. Bloch and W. W. Hastings, *Plastics Materials in Surgery,* 2nd edition, Charles C Thomas, Springfield, Ill., 1972.
J. W. Boretos, *Concise Guide to Biomedical Polymers,* Charles C Thomas, Springfield, Ill., 1973.
P. J. Flory, *Principles of Polymer Chemistry,* Cornell University Press, Ithaca, N.Y., 1953.
H. P. Gregor (ed.), *Biomedical Applications of Polymers,* Plenum Press, New York, 1975.
R. L. Kronenthal and Z. Oser, (eds.), *Polymers in Medicine and Surgery,* Plenum Press, New York, 1975.
H. Lee and K. Neville, *Handbook of Biomedical Plastics,* Pasadena Technology Press, Pasadena, Calif., 1971.
D. J. Lyman, "Biomedical Polymers," *Rev. Macromol. Chem., 1,* 355, 1966.
P. Meares, *Polymers: Structure and Bulk Properties,* D. Van Nostrand Co., London, 1965.
J. M. Schulz, *Polymer Materials Science,* Prentice-Hall, Englewood Cliffs, N.J., 1974.

CHAPTER 7

STRUCTURE–PROPERTY RELATIONSHIPS OF BIOLOGICAL MATERIALS

The major difference between biological materials and biomaterials (implants) is viability. Other equally important differences distinguish living materials from artificial replacements. First, most biological materials are continuously bathed with a solution of water. Exceptions are the specialized surface layers of skin, hair, nails, hooves, and the enamel of teeth. Second, most biological materials can be considered as composites.

Basically, biological tissues consist of a vast network of intertwining fibers with polysaccharide ground substance immersed in a pool of ionic fluid. Attached to the fibers are cells whose responsibility is nutrition of the fibers and ground substances that carry out the physical function. The ground substances probably have definite structural organizations and are not completely analogous to solute suspended in a solution. Physically, ground substance behaves as a glue, lubricant, and shock absorber in various tissues.

The structure and hence the properties of a given biological material are dependent on the chemical and physical nature of the components present and their relative amount. For example, nervous tissue consists almost entirely of cells, while bone is composed of collagenous fibers and calcium phosphate minerals with minute quantities of cells and ground substance (as a glue).

Understanding the exact role played by a biological tissue and its interrelationships with the functions of the entire living organism is essential if biomaterials are to be used intelligently. Thus, designing an artificial blood vessel prosthesis requires understanding not only the blood vessel wall property–structure relationship itself but also the systemic function. This is because the artery is not only a conduit for blood but is a component of a larger system, including a pump (heart) and an oxygenator (lung).

7.1. STRUCTURE OF PROTEINS AND POLYSACCHARIDES

7.1.1. Proteins

Like polymeric materials, proteins are made of monomers called peptides [equation (6-3)]:

$$-\underset{H}{\overset{H}{\underset{|}{C}}}\left(\underset{}{\overset{O}{\overset{\|}{C}}}-\underset{H}{\overset{H}{\underset{|}{N}}}-\underset{}{\overset{H}{\underset{|}{C}}}\right)_n- \qquad (7\text{-}1)$$

The peptides are in turn amides formed by interaction between amino and carboxyl groups of amino acids; and the basic chemical formula is

$$HN-\underset{R}{\overset{H}{\underset{|}{C}}}-\overset{O}{\overset{\|}{C}}-OH \qquad (7\text{-}2)$$

where R is a side group. Depending on the side group the molecular structure changes drastically. The simplest side group is hydrogen (H), which will form glycine. The geometry of the peptide is shown in Figure 7-1a, where the hypothetical flat sheet structure is shown. The structure has a repeat distance of 7.2 Å and the side groups (R) are crowded. This crowding of side groups makes the flat structure impossible except for the H side group, i.e., polyglycine. If the side chains are larger still, then the resulting structure is a helix, where the H bonds occur between different parts of the same chain and hold the helix together as shown in Figure 7-1.

7.1.1.1. Collagen. One of the basic constituents of proteins is collagen, which has the general amino acid sequence -Gly-Pro-Hypro-Gly-X- with a triple α helix. It has a high proportion of proline and hydroxyproline, as given in Table 7-1. Because the contents of hydroxyproline are unique in collagen (elastin contains minute amount), the collagen content in a given tissue is easily determined by assaying the hydroxyproline.

BIOLOGICAL MATERIALS

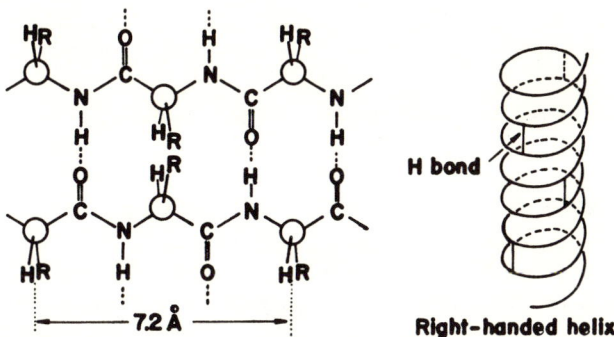

Figure 7-1. Left: Hypothetical flat sheet structure of a protein. Right: Helical arrangement of a protein chain.

Table 7-1. Amino Acid Content of Collagen[a]

Amino acid (a.a.) and component	Content (mol/100 mol a.a.)
Gly	31.4–33.8
Pro	11.7–13.8
Hypro	9.4–10.2
Acid polar a.a. (Asp, Glu, Asparagine)	11.5–12.5
Basic polar a.a. (Lys, Arg, His)	8.5–8.9
Other a.a.	Residue

[a] By permission from M. Chvapil, *Physiology of Connective Tissue*, p. 121, Butterworths, London, 1967.

Three left-handed α-helical peptide chains are coiled together to give a right-handed coiled helix with a periodicity of 28.6 Å. This triple helix super helix is the molecular basis of tropocollagen, precursor of collagen. The three chains are held together strongly to each other by H bonds between glycine residues and between the hydroxyl (—OH) groups of hydroxyproline. In addition, there are cross-linkages via lysine among the α helixes. When the cross-linkage between α helixes are done by two and three α helixes they are called α and γ chains, respectively. The labeling system of the (tropo) collagen fractions can be expressed as

$$\begin{aligned} \alpha_1 + \alpha_2 &\to \beta_{12} \\ \alpha_1 + \alpha_3 &\to \beta_{12} \\ \alpha_1 + \alpha_3 &\to \beta_{23} \\ \alpha_1 + \beta_{12} &\to \gamma_{112}, \text{ etc.} \end{aligned} \quad (7\text{-}3)$$

Because the molecular weight of an α fraction of the tropocollagen is 120,000 g/mol, the β and γ chains will have 240,000 and 360,000 g/mol, respectively.

Depending on the source of tissues the collagen chains are different. Bone and skin collagen contains two α_1 (type I) and β_2 chains; the cartilage is composed of three identical chains called α_1 (type II) chains but has a different primary structure. On the other hand, blood vessel walls and dermis have α_1 (type III) chains, which contain cysteine that can be cross-linked within molecules through disulfide bonds.

The primary factors stabilizing the collagen molecules are invariably related to the linkage between the α helixes. These factors are H bonding between the C=O and NH groups, and ionic bonding between the side groups of polar amino acids and the interchain cross-links. One of the secondary factors affecting the stability of collagen is steric rigidity, which is related to the high pyrolidine content.

The side groups of the amino acids of collagen are highly nonpolar in character and hence hydrophobic; thereby the chains avoid contact with water molecules and seek the greatest number of contacts with the nonpolar chains of other amino acids. If the hydrophobic contact is destroyed by a solution (e.g., urea) then the characteristic structure is lost, resulting in microscopic changes such as shrinkage of collagen fibers. The same effect can be achieved by simply warming the collagen solution. Another factor affecting the stability of the collagen is water molecules incorporated into the intra- and interchain structure. If the water content is lowered the structural stability decreases, and if dehydrated completely (lyophilized) then the solubility also decreases (so-called *in vitro* aging of collagen).

It is known that the acid mucopolysaccharides also affect the stability of collagen fibers by mutual interaction or by forming mucopolysaccharide-protein complexes. It is believed that the water molecules affect the polar region of the chains as shown schematically in Figure 7-2. Note that the dried collagen is more disoriented than the wet one, indicating that the water molecules fill up the intervening space and allow the backbone helices of the fibrils to be arranged in parallel.

7.1.1.2. **Elastin.** Elastin is another structural protein found in relatively large amount in such elastic tissues as ligament, aortic wall, and skin. The chemical composition of elastin is somewhat different from that of collagen.

The high elasticity of the elastin is believed to result from the cross-linking of lysine residues into desmosine, isodesmosine, and lysinonorleucine, as shown in Figure 7-3. The formation of desmosine and isodesmosine is only possible by the presence of Cu^{2+} and lysyl oxidase enzyme; hence deficiency of copper in the diet may result in non-cross-linked elas-

Figure 7-2. Schematic representative of collagenous protofibrils in dry (left) and wet states. (Redrawn from R. S. Bear, *Advan. Protein Chem.*, 7, 149, 1952, courtesy of Academic Press, New York.)

tin. This in turn will result in viscous nonelastic tissue similar to the linear polymers; the tissue loses its rubberlike elasticity, which can lead to rupture of the aortic walls.

The elastin is very stable at high temperature and chemicals because of the very low content of polar side groups (hydroxy and ionizable groups). The specific staining of elastin in tissues by lipophilic stains such as Weigert's resorcin-fuchsin results from the same reason. Like collagen, elastin contains high percentages of alanine, proline, and glycine as well as a high proportion of amino acids with aliphatic side chains, such as valine (6 times that of collagen). It also lacks all the basic and acidic amino acids so that it has very few ionizable groups. The most abundant of these, glutamic acid, occurs only one-sixth as often as in collagen. Aspar-

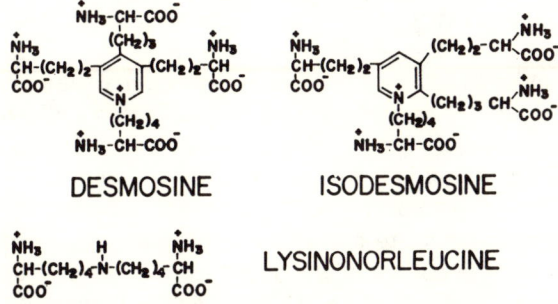

Figure 7-3. Structures of desmosine, isodesmosine, and lysinonorleucine.

Table 7-2. Overall Composition of Elastin Protein (Residues/1000)[a]

Amino acids	Elastin	Microfibril	Tropoelastin
Glycine	324	110	334
Hydroxyproline	26	151	39
Cationic residues (Asp, Glu)	21	228	21
Anionic residues (His, Lys, Arg)	13	105	55
Nonpolar residues (Pro, Ala, Val, Met, Leu, Ileu, Phe, Try)	595	356	541
Cys/2	4	48	0

[a] From various sources.

tic acid, lysine, and histidine are all below 2 residues per 1000 in mature elastin.

Three major entities of elastin have been separated: tropoelastin, elastin, and microfibrils. The composition of these components are given in Table 7-2. The microfibrillar phase, which is largely oriented crystalline structure, may be 5–10% of the total elastin and the rest is amorphous.

7.1.2. Polysaccharides

Polysaccharides exist in tissues as highly viscous materials that interact readily with proteins, including collagen-resulting glycosaminoglycans or proteoglycans and readily bind both water and cations. The compounds also exist at physiological concentrations as viscoelastic gels, not as viscous sols. All these polysaccharides consist of repeating disaccharide units polymerized to unbranched macromolecules, as shown in Figure 7-4.

7.1.2.1. Hyaluronic Acid and Chondroitin. Hyaluronic acid is made of residues of N-acetylglucosamine and D-glucuronic acid but lacks sulfate residues. The animal hyaluronic acid contains protein component (0.35 w/o or more) and is believed to be chemically bound to at least one protein or peptide that cannot be removed. This, in turn, will result in proteoglycan molecules, which may behave differently from the pure polysaccharides. Hyaluronic acid is found in the vitreous humor of the eye, synovial fluid, skin, umbilical cord, and aortic walls.

Chondroitin is similar to hyaluronic acid in its structure and properties; it is found in the cornea of the eyes.

7.1.2.2. Chondroitin Sulfate. This is a sulfated mucopolysaccharide and resists hyaluronidase. It has three isomers, as shown in Figure 7-4.

BIOLOGICAL MATERIALS

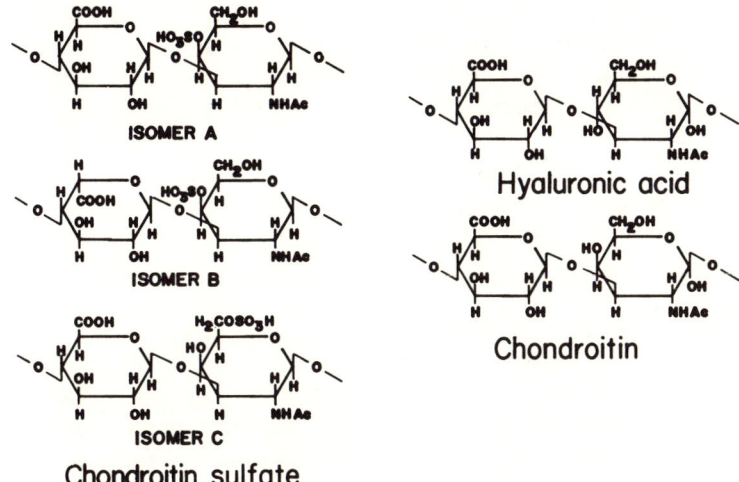

Figure 7-4. Structures of hyaluronic acid, chondroitin, and chondroitin sulfates.

Isomer A (chondroitin 4-sulfate) is found in cartilage, bones, and cornea, while isomer C (chondroitin 6-sulfate) can be isolated from cartilage, umbilical cord, and tendon. Isomer B (dermatan sulfate) is found in skin and lung and is resistant to testicular hyaluronidase.

The chondroitin sulfate chains in connective tissues are bound covalently to a polypeptide backbone through their reducing ends. Figure 7-5 shows a proposed macromolecular structure of protein–polysaccharides

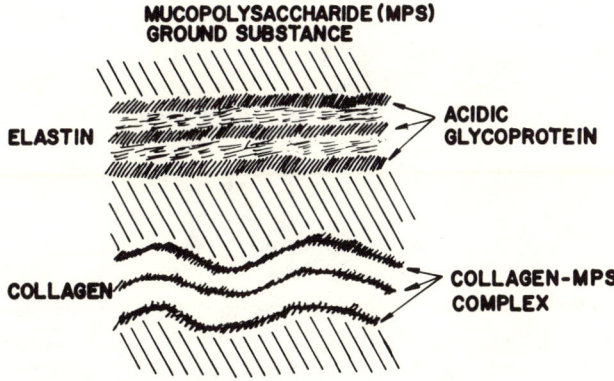

Figure 7-5. A schematic arrangement of mucopolysaccharides–protein molecules in connective tissues.

from which one can imagine the viscoelastic properties of the ground substance. These complexes of protein and polysaccharides play an important role of the physical behavior of connective tissues, either as lubricating agents between tissues (like joints) or between elastin and collagen microfibrils.

Similar structural stabilizing function of the mucopolysacharides is shown in Figure 7-5, where the collagen fibrils are stabilized by the mucopolysaccharides–protein interactions.

7.2. STRUCTURE–PROPERTY RELATIONSHIP OF TISSUES

Understanding the structure–property relationship of various tissues is essential for implantation. The property measurements of any tissues are confronted with many of the following limitations and variations:

1. Limited sample size.
2. The original structure can be changed during sample collection or preparation.
3. Inhomogeneity.
4. Cannot be frozen or homogenized without altering its structure or properties.
5. The complex nature of the tissues makes it difficult to obtain fundamental physical parameters.
6. *In vitro* and *in vivo* property measurements are sometimes difficult, if not impossible, to correlate.

The main objective of studying the property–structure relation is to replace or revitalize the function of a given tissue or organ. Thus one must always ask what kind of functions of the tissue or organ under study are being performed *in vivo* and how can the properties be best represented with as few parameters as possible. Each individual tissue structure and property relationship can be studied in this manner.

7.2.1. Collagen-Rich and Mineralized Tissues

7.2.1.1. Bone and Teeth.
Bone and teeth are mineralized tissues whose primary function is "load-carrying." Teeth are in more extraordinary physiological circumstances because their function is carried out in direct contact with *ex vivo* substances, whereas most bone functions are carried out inside the body in conjunction with muscle and tendon.

Bone is composed of 40% organic material, of which 90–96% is collagen and the rest is mineral. The major subphase of the mineral consists of submicroscopic crystals of an apatite of calcium and phosphate resembling hydroxyapaptite crystal structure [$Ca_{10}(PO_4)_6(OH)_2$]. There are other mineral ions, such as citrate ($C_6H_5O_7^{4-}$), carbonate (CO_3^{2-}), fluoride (F^-), and hydroxyl ions (OH^-), which may give some other subtle differences in microstructural features of the bone. The apatite crystals are formed as slender needles 200–400 Å long by 15–30 Å thick in the collagen fiber matrix. These mineral–containing fibrils are arranged into lamellar sheets (3–7 μm thick) and sheets run helically with respect to the long axis of cylindrical osteon (sometimes called the Haversian system), as shown in Figure 7-6. The osteon is made up of from 4 to 20 lamellae arranged in concentric rings around the Haversian canal. Between these osteons the interstitial systems are dispersed and the limits of the Haversian and interstitial systems are sharply divided by the cementing line. The metabollic substances can be transported by the intercommunicating systems of canaliculi, lacunae, and Volkman's canal, which are connected with marrow cavity. The external and internal surface of the bone is called periosteum and endosteum, respectively, and both have osteogenic properties.

Long bones such as the femur are usually made of spongy and compact bone. The spongy bone consists of three-dimensional branching of

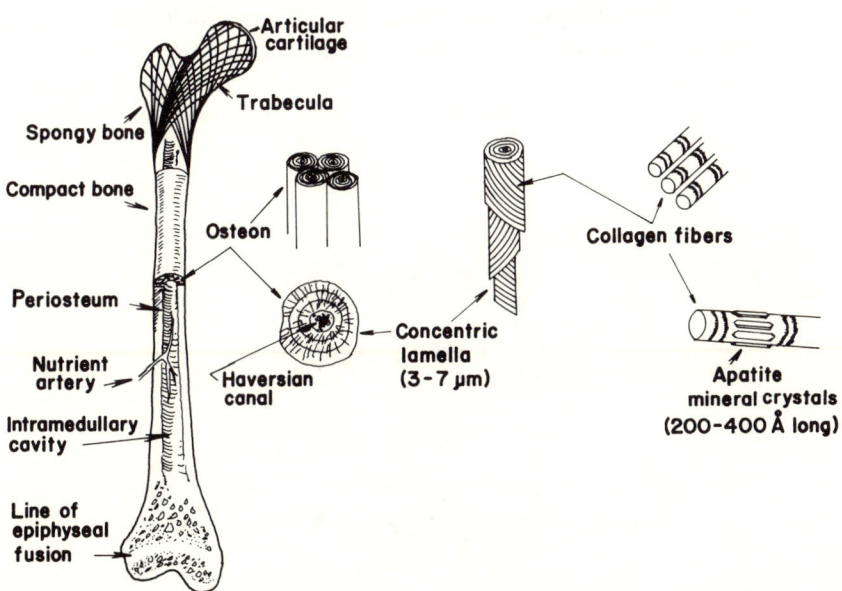

Figure 7-6. Micro- and macrostructure of a bone.

bony or trabeculae interdispersed by bone marrow. The spongy bone changes gradually into compact bone toward the middle of the bone.

There are two types of teeth, deciduous (or primary) and permanent; the latter is more important for understanding implantation. All teeth are made of two portions, the crown and the root, demarcated by the gingiva (gum). The root is placed in a socket, called the alveolus, in the maxillary (upper) or mandibular (lower) bone. A sagittal cross section of a permanent tooth is shown in Figure 7-7 to illustrate various structural features.

Tooth enamel is the hardest substance found in the body and consists almost entirely of calcium phosphate salts (97 percent) in the form of large apatite crystals. Dentin is another mineralized tissue whose distribution of organic matrix and mineral is similar to the regular bone and its physical properties. The collagen matrix of the dentin might have a somewhat different molecular structure than normal bone; it is more cross-linked, i.e., it has more β and γ chains of collagen than in other tissues, resulting in less swelling effect. Dentinal tubules (3–5 μm diameter) radiate from the pulp cavity toward the periphery and penetrate every part of dentin. Collagen fibrils (2–4 μm diameter) are filled inside the dentinal tubules in longitudinal direction and the interface is cemented by protein–polysaccharides.

Cementum covers most of the root of the tooth with coarsely fibrillated bone substance but is devoid of canaliculi, Haversian systems, and blood vessels. The pulp occupies the cavity and contains thin collagenous fibers running in all directions and not aggregated into bundles. The ground substances, nerve cells, blood vessels, etc., are also contained in the pulp. The periodontal membrane anchors the root firmly into the

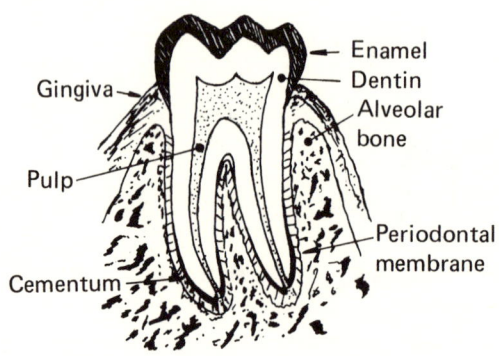

Figure 7-7. Schematic diagram of a tooth.

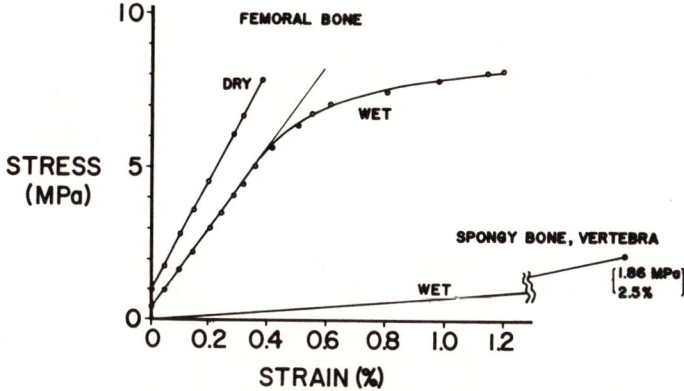

Figure 7-8. Compressive stress and strain behavior of a dry and a wet bone. (Redrawn by permission from G. H. Kummer, *Biomechanics, Its Foundations and Objectives,* ed. Y. C. Fung et al., Prentice-Hall, Englewood Cliffs, N.J., 1972.)

alveolar bone and it is made mostly of collagenous fibers plus glycoproteins (protein–polysaccharides complex).

As with most other biological materials, the property of bone depends largely on the humidity, mode of applied load (compression or tensile, static or dynamic, etc.), and direction of applied load onto the sample. Therefore, one usually studies the effect of the previously mentioned factors and correlates the results with structural features. For brevity I will follow the same practice. The effect of drying of the bone can be seen easily from Figure 7-8, where the dry sample shows a higher modulus of elasticity and compression strength, and lower toughness and fracture strain. Thus the wet bone, which is similar to *in vivo* characteristics, can absorb more energy and elongate more before fracture.

As with any other brittle material, there is a large variation in reported values of bone strength measurements. However, unlike brittle materials, the bone (and dentin) undergoes creep and stress relaxation similar to such viscoelastic materials as polymers. This can be interpreted to mean that the major network of structure is composed of highly cross-linked collagen and the apatite crystals are distributed discretely, as electron microscopic pictures reveal. It is also interesting that for a given bone, the strength is higher in the direction of physiological loading (e.g., the strength is higher in longitudinal direction than the radial in humerus), which can be viewed as a manifestation that Wolff's law prevails. Thus the strength of a male's bone is higher than that of a female's not because of sex but because men load their bones more rigorously by being heavier and carrying more loads.

Bone is also constructed very efficiently from a structural mechanics viewpoint because a given material can best be utilized by making it in a tubular shape, which has the same bending strength as a solidly filled tube but requires much less material (see Section 2.1). Studies of the compressive strength and density showed that only 40–42% of the difference in strength can be attributed to the density difference. This illustrates that mechanical strength is more related to structural integrity than is the amount of mineral phase present.

Although the continuum mechanical formulation of the stress–strain behavior of bone in terms of a simple equation (such as equation 2-1) is highly informative, it does not reveal the relative contribution of two major structural components (i.e., organic and mineral components) to overall mechanical behavior. The mechanical behavior of bone can be interpreted as a conventional two-phase system similar to steel-reinforced concrete. Therefore, if one assumes that the load is independently borne by the two components (collagen and mineral, hydroxyapatite), then the total load (P_t) is borne by mineral (P_m) and collagen (P_c) together,

$$P_t = P_m + P_c \qquad (7\text{-}4)$$

Since

$$\sigma = P/A = E\epsilon \quad \text{then} \quad P_c = A_c \times E_c \times \epsilon_c \qquad (7\text{-}5)$$

where A, E, and ϵ are area, modulus, and strain, respectively. The strain of collagen can be assumed to be equal to that of mineral, i.e., $\epsilon_c = \epsilon_m$; thus

$$P_c = P_t \frac{A_c E_c}{A_m E_m} \qquad (7\text{-}6)$$

Therefore,

$$P_m = \frac{P_t A_m E_m}{A_m E_m + A_c E_c} \qquad (7\text{-}7)$$

As a simple example, assume that a load of 1000 newtons is acting on a cross-sectional area of 10 mm²; hence $P_t = 100$ MPa. Because the modulus of elasticity of collagen and bone is about 0.1 GPa and 17 GPa, respectively (see Table 7-3), and the volume fraction of each component is about the same, the fraction of load borne by collagen and mineral phase becomes 0.006 and

Table 7-3. Properties of Bone[a]

Tissue	Direction of test	Modulus of elasticity (GPa)	Tensile strength (MPa)	Compressive strength (MPa)
Leg bone				
Femur	Longitudinal	17.2	121	167
Tibia	Longitudinal	18.1	140	159
Fibula	Longitudinal	18.6	146	123
Arm bone				
Humerus	Longitudinal	17.2	130	132
Radius	Longitudinal	18.6	149	114
Ulna	Longitudinal	18.0	148	117
Vertebra				
Cervical	Longitudinal	0.23	3.1	10
Lumbar	Longitudinal	0.16	3.7	5
Spongy bone		0.09	1.2	1.9
Skull	Tangential	—	25	
	Radial	—		−97

[a] From H. Yamada, *Strength of Biological Materials,* chapter 3, Williams & Wilkins Co., Baltimore, 1970.

0.994. This indicates that most of the load is carried by mineral phase at a normal loading condition. Actually, the strength of the demineralized bone is about 5–10% of the whole bone.

Some of the physical properties of the teeth are given in Table 7-4. As can be expected, the strength is the highest for enamel, and dentin is between bone and enamel. The thermal expansion and conductivity are important factors to consider because filling materials should match with them. The thermal expansion and conductivity are higher for enamel than for dentin, as shown in Table 7-4.

Table 7-4. Physical Properties of Teeth[a]

Tissue	Density (g/cm^3)	Modulus of elasticity (GPa)	Compressive strength (MPa)	Coefficient of thermal expansion (cm/cm/°C)	Thermal conductivity (W/m°K)
Enamel	2.2	48	241	11.4×10^{-6}	0.82
Dentin	1.9	13.8	138	8.3×10^{-6}	0.59

[a] From various sources.

Example 7-1

A bone sample with a square cross section of 0.3 cm in one side has a 0.5-cm gage mark. The following data were obtained:

Load (N)	Gage length (cm)
0	0.5000
250	0.5013
500	0.5025
750	0.5038
1000	0.5063
1250	0.5090 (broke)

a. Draw the stress–strain curve.
b. What is the yield strength?
c. What is the breaking strength?
d. What is the modulus of elasticity?

Answers
a.

Stress (MPa)	Strain (%)
0	0
27.8	0.26
55.6	0.50
83.3	0.76
111.1	1.26
138.9	1.80

b. From the curve plotted of stress versus strain, the yield stress is <u>80 MPa</u>.

c. <u>138.9 MPa</u>.

d. $E = \dfrac{27.8 \text{ MPa}}{0.26 \times 10^{-2}} = \underline{10.7 \text{ GPa}}$

(about half the reported value).

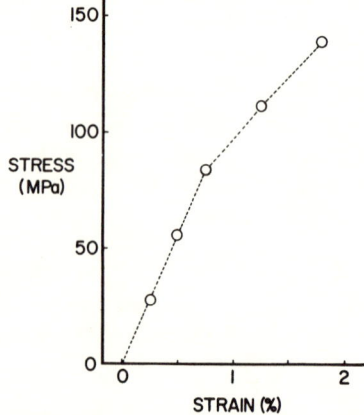

7.2.1.2. Tendon. The main physiological function of tendon is to transmit tensile forces between muscle and bone. Thus the tendon is made of highly oriented collagen fibers (30 w/o wet; 75 w/o dry) in a

Table 7-5. Elastic Properties of Some Nonmineralized Human Tissues[a]

Substance	Tensile strength (MPa)	Ultimate elongation (%)
Skin	7.6	78
Tendon	53	9.4
Elastic cartilage	3	30
Heart valves (aortic)		
radial	0.45	15.3
circumferential	2.6	10.0
Aorta		
transverse	1.1	77
longitudinal	0.07	81
Cardiac muscle	0.11	63.8

[a] From various sources.

longitudinal direction. It contains 58–70 w/o water in body and few elastic fibers (0.8 w/o wet; 2 w/o dry). Where lubrication is needed the tendon is constrained within a collagenous sheath to facilitate rapid movement.

Tendon is one of the least extensible among the nonmineralized tissues and the highest in tensile strength, as shown in Table 7-5. However, the tensile strength of 53 MPa is orders of magnitude lower than that of individual collagen fibers (3000 MPa from Table 7-6), even though the tendon is made of highly oriented collagenous fibrils. This low strength is largely caused by the imperfection of fibril organization, and the ends of the individual fibrils may act like defect regions as in textile fibers, hence reducing tendon strength considerably.

7.2.1.3. Cartilage. The physiological function of cartilage can be divided into two parts; maintenance of shape and providing bearing surfaces at joints. The chemical composition of cartilage is similar to tendon, but the collagen fibers are not uniformly oriented as in tendon. Cartilage contains very large and diffuse polysaccharide–protein molecules that form a gel in which the collagen molecules could be entangled (cf. Figure 7-5). They

Table 7-6. Elastic Properties of Elastic and Collagen Fibers[a]

Substance	Modulus of elasticity (MPa; 100% elongation)	Tensile strength (MPa)	Ultimate elongation (%)
Elastic fibers	0.3	1	100
Collagen	100	3000	50

[a] From various sources.

can affect the mechanical properties of the cartilage by hindering the movements of fluid through the interstices of the collagenous matrix network. The supportive cartilage can be found in ear, tip of nose, rings around the trachea, and ends of ribs joining to the bony sternum.

The joint cartilage has a very low coefficient of friction (0.01). The lubricating function is carried out in conjunction with mucopolysaccharides, especially chondroitin sulfates, as discussed previously in Section 7.1.2.2. The modulus of elasticity (10.3–20.7 MPa) and tensile strength (3.4 MPa) are quite low. However, wherever high stress is required, the cartilage is replaced by purely collagenous tissue.

7.2.1.4. Skin. The skin is composed of epidermis and dermis. The latter is mostly responsible for the mechanical strength while the epidermis is primarily important for water conservation. The skin performs many other obvious functions such as a barrier for heat transfer, electrical shock, bacteria, or radiation.

The skin is composed largely of collagen (75 w/o dry) with ground substances of mucopolysaccharides (20 w/o dry) for flexibility, as in tendon and cartilage. The skin is structurally anisotropic; hence depending on the direction of testing, the response of load can be different, as shown in Figure 7-9. This anisotropy is the reason behind the Langer's line, which follows the least resistant path of the dermis for easier surgery and better wound healing.

Another feature of the stress–strain curve of the skin is its extensibility with a small load despite its high collagen content, which is largely a result of the highly random state of the dermal fibers. On stretching, the

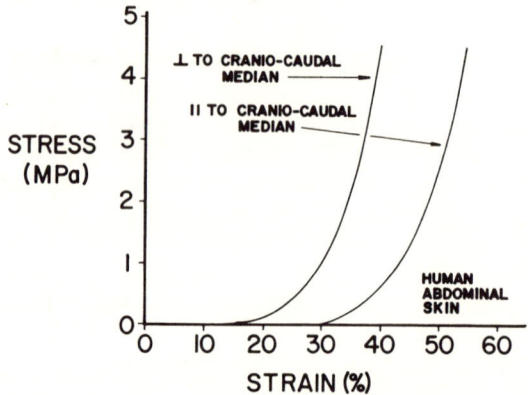

Figure 7-9. Stress–strain curves of human abdominal skin. (After C. H. Daly, *The Biomedical Characteristics of Human Skin,* Ph.D. Thesis, University of Strathclyde, Scotland, 1966.)

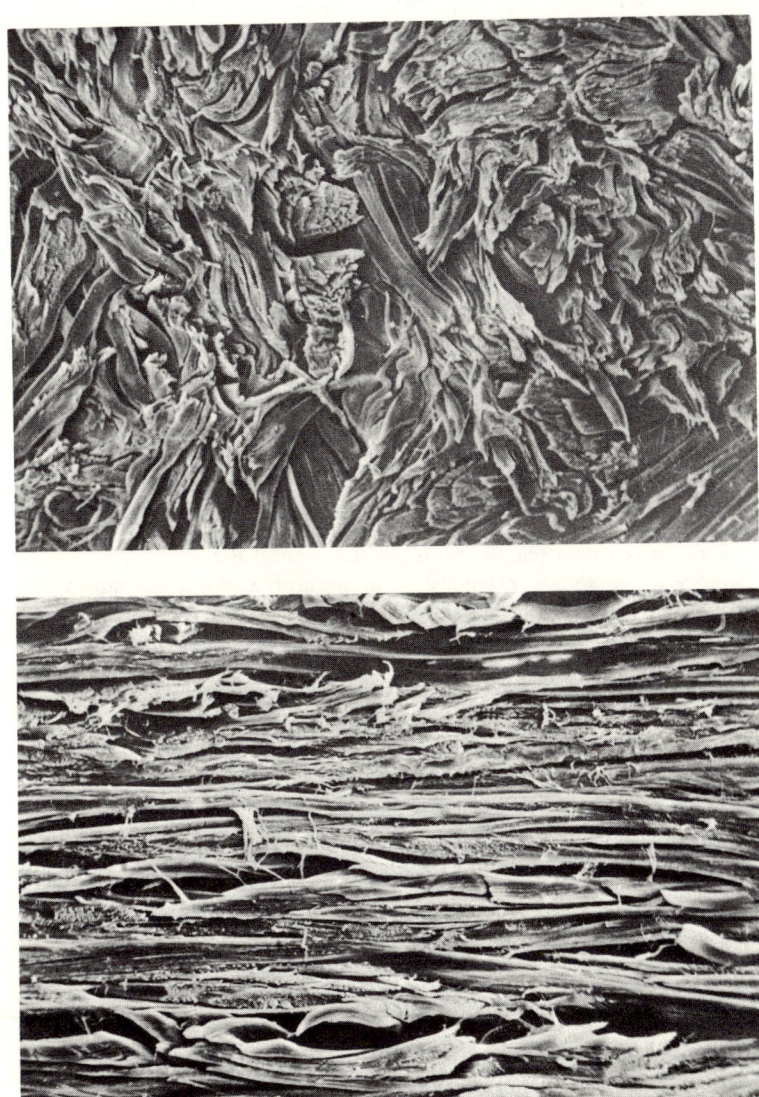

Figure 7-10. Scanning electron microscopic pictures of a dermal skin before and after stretching. Stretch direction is horizontal, 400x. [From G. L. Wilkes, I. A. Brown, and R. H. Wildnauer, *CRC Crit. Rev. Bioeng.*, *1* (4), 459, 1973.]

fibers align each with other and resist further extension, as can be seen in Figure 7-10. When the skin is highly stretched, the modulus of elasticity at these high-extension portions approaches that of tendon.

Example 7-2

The curve at the right was obtained by tensile testing of canine skin. The skin specimen was cut by using a stamping machine of width 4 mm. The thickness of skin is about 3 mm and the length of the sample between the grips is 20 mm.

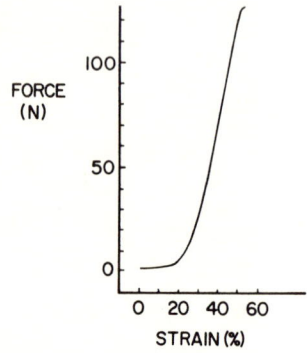

Force versus strain of canine dorsal skin.

a. What is the tensile strength and fracture strain of the specimen?
b. What are the moduli of elasticity in the initial region and secondary region?
c. What is the toughness of the skin?

Answers

a. $$\sigma = \frac{129 \text{ N}}{3 \times 4 \times 10^{-6} \text{ m}^2} = \underline{10.75 \text{ MPa}}$$

b. $$E_i = \frac{0.186}{0.55} = \underline{0.34 \text{ MPa}}$$
$$E_s = \frac{11.5 - 0.1}{0.55 - 0.25} = \underline{38 \text{ MPa}}$$

c. Area under the force–strain curve is approximated by a triangle: $(0.56 - 0.22) \times 129/2 = 21.93$ N · m/m. Because the cross-section area is 12×10^{-6} m², the toughness becomes

$$\frac{21.93}{12 \times 10^{-6}} \frac{\text{N} \cdot \text{m}}{\text{m}^3} = \underline{1.83 \times 10^{-6} \frac{\text{N} \cdot \text{m}}{\text{m}^3}}$$

7.2.2. Elastic Tissues

7.2.2.1. Blood Vessel Walls. The blood vessel has three distinct layers in cross section (Fig. 7-11): (1) intima, whose structural elements are oriented (longitudinally); (2) media, which is the thickest layer of the

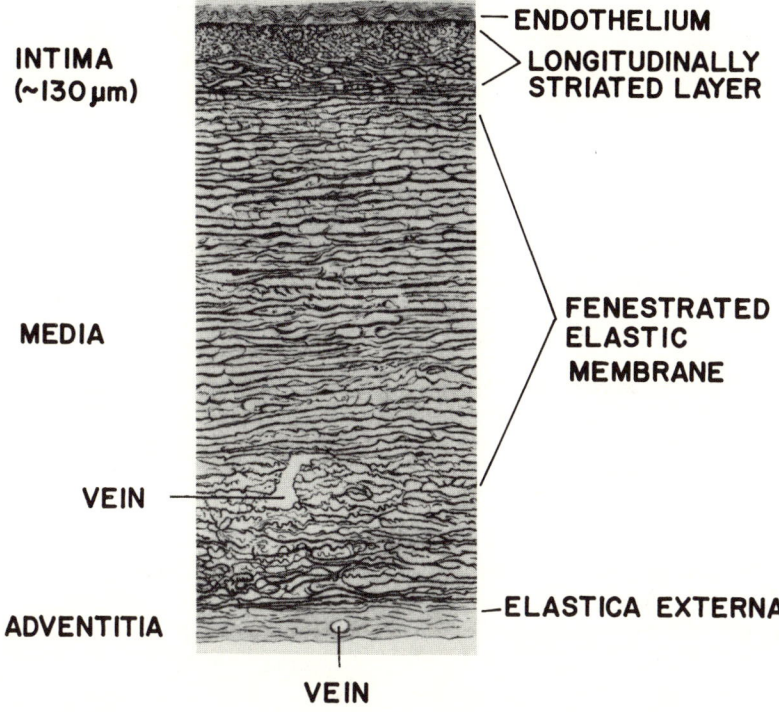

Figure 7-11. Structure of a blood vessel wall. (From W. Bloom and D. W. Fawcett, *A Textbook of Histology*, 9th ed., p. 378, W. B. Saunders Co., Philadelphia, 1968, by permission from the publisher.)

wall (and its components are arranged circumferentially); and (3) adventitia, which firmly connects the vessels to the matrix via facia. The bonding between intima and media is formed by the internal elastic membrane (*elastica interna*) predominant in arteries of medium size. Between the media and adventitia, a thinner external elastic membrane (*elastica externa*) can be found. The smooth muscle cells are found between adjacent elastic lamellae in helical array.

Because of the anisotropy of the blood vessel arrangement and its tubal structure the intrinsic properties are not well defined. Early studies of mechanical properties of blood vessels were done by pressurizing with saline solution into a segment of the vessel after closing the branches, and recording the pressure and diameter/length changes as shown in Figure 7-12. However, the composition of the vessel walls changes along the length

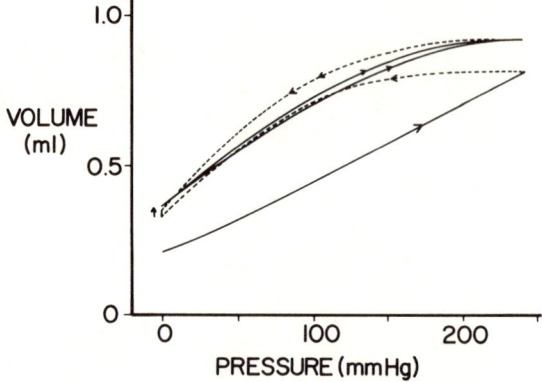

Figure 7-12. Repeated volume versus pressure tests of excised canine femoral artery. (Redrawn from D. H. Bergel, *J. Physiol.*, **156**, 445, 1961, courtesy of the Syndics of the Cambridge University Press.)

of the wall and hence their physical properties also change (Fig. 7-13). Another complicating factor is the existence of smooth muscle, which is associated with arterial blood pressure regulation.

Table 7-7 shows the mean pressure of the various blood vessels and the approximate tension developed at normal pressure calculated by using

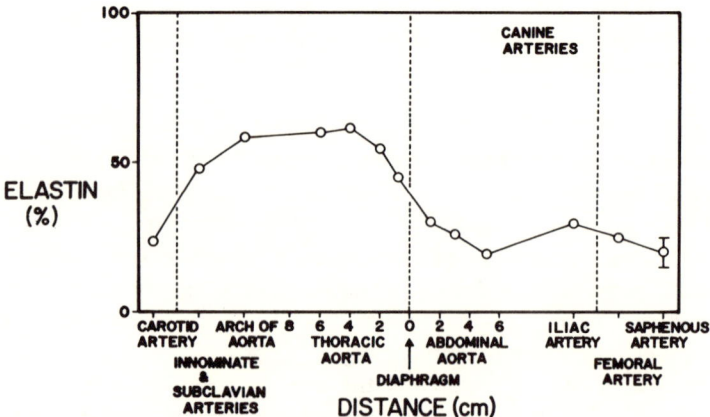

Figure 7-13. Variation of elastin percent per elastin plus collagen along the major arterial tree. (Redrawn from R. D. Harkness, "Mechanical Properties of Collagenous Tissues," in *Treatise on Collagen,* ed. B. S. Gould, chapter 6, vol. 2, part A, Academic Press, New York, 1968, by permission from the publisher.)

BIOLOGICAL MATERIALS

Table 7-7. Wall Tension and Pressure Relationship of Blood Vessels of Various Sizes[a]

Vessel	Mean pressure (mm Hg)	Internal pressure (dynes/cm^2)	Radius	Wall tension (dynes/cm)
Aorta, large artery	100	1.5×10^5	1.3 cm <	170,000
Small artery	90	1.2×10^5	0.5 cm	60,000
Arteriole	60	8×10^4	62 μm–0.15 mm	500–1200
Capillaries	30	4×10^4	4 μm	16
Venules	20	2.6×10^4	10 μm	26
Veins	15	2×10^4	2 mm	400
Vena cava	10	1.3×10^4	1.6 cm	21,000

[a] By permission from A. C. Burton, *Physiology and Biophysics of the Circulation*, chapter 7, Year Book Medical Publishers, Chicago, 1965.

the Laplace equation

$$T = P \cdot r \qquad (7\text{-}8)$$

where T is wall tension, P is internal pressure, and r is the radius of the vessel. This theory is usually applied to uniform, thin, isotropic walls in the absence of longitudinal tension. It is evident that none of the requirements can be met strictly by blood vessel walls.

7.2.2.2. Ligament. Like tendon, ligament is largely composed of collagen, elastin, and ground substances but it contains a much higher proportion of elastin (60–70 w/o dry). This in turn results in much higher elongation before fracture and at low strain it can be recovered fully without a permanent set. This is particularly suitable for maintaining the head of a grazing animal such as an ox (ligamentum nuchae) because it requires high extension and retraction without permanent damage to the tissue. This rubberlike elasticity is found true up to 50 percent elongation, as shown in Figure 7-14. However, if the collagen is removed from the ligament by enzyme or autoclaving (17 w/o dry), it behaves like an elastomer, with up to 100 percent elongation (Fig. 7-14). From this result, when two components are pulled together in a simple model, as shown in Figure 7-15, the elastic fibers (elastin) are stretched first, followed by the much stronger collagen fibers. With a little more imagination the distribution of fiber length can be envisioned; thus the response can be smoother than the two-fiber system shown in Figure 7-14. This kind of study relating the microstructure to the macrobehavior allows a fuller understanding of the nature of interaction between components of a connective tissue.

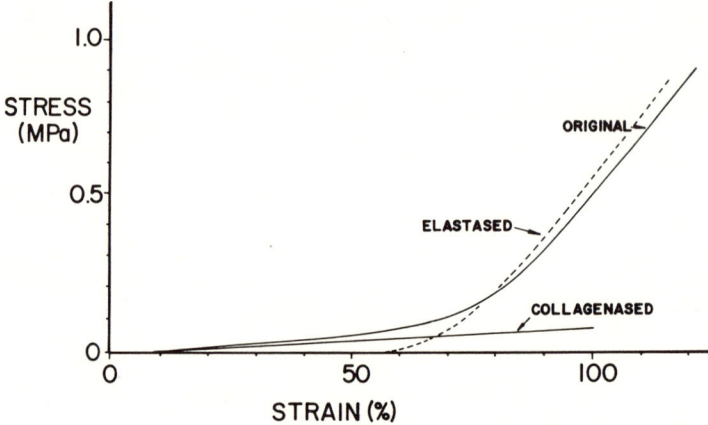

Figure 7-14. Stress–strain curves of bovine ligamentum nuchae after elastase and collagenase treatments. (Modified from A. S. Hoffman, L. A. Grande, P. Gibson, J. B. Park, C. H. Daly, and R. Ross, "Preliminary Studies on Mechanochemical–Structure Relationships in Connective Tissues Using Enzymolysis Techniques," in *Perspectives of Biomedical Engineering,* ed. R. M. Kenedi, p. 173, University Park Press, Baltimore, 1973.)

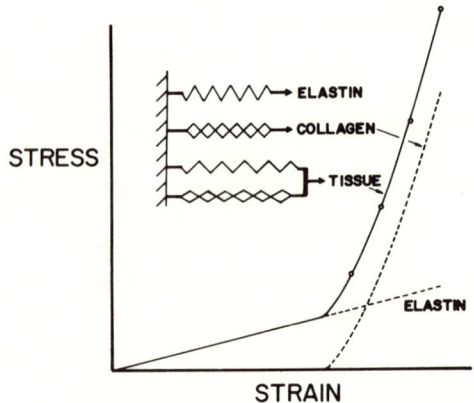

Figure 7-15. The structure–property relationship of connective tissues. The elastin is represented by a spring or an elastic rubber and the collagen by a loosely knitted fabric.

BIOLOGICAL MATERIALS

Example 7-3

Ligamentum nuchae is made of elastin and collagen (other elements do not contribute toward its mechanical strength). The relative amounts excluding water are 70 w/o elastin, 25 w/o collagen, and 5 w/o others; they are homogeneously distributed.

a. Calculate the percent contribution of the elastin toward the total strength assuming elastic behavior (use Table 7-6).
b. Compare the result with Figure 7-14.

Answers

a. From equation (7-7),

$$P_c = \frac{P_t A_c E_c}{A_c E_c + A_e E_e}$$ where P_c and P_t are loads borne by collagen and total tissue.

$$\frac{P_c}{P_t} = \frac{0.7 A_t \cdot 100}{0.7 A_t \cdot 100 + 0.25 A_t \cdot 0.3} \qquad \begin{array}{l} A_c = 0.7 A_t \\ A_e = 0.25 A_t \end{array}$$

$$= \frac{70}{70 + 0.075}$$

$$= \underline{0.9989} \text{ (99.89% of loads borne by collagen)}$$

b. The collagen contributes to the total load at high strain (>55%), but elastin contributes a large proportion at lower strains. However, the contribution of collagen at the low strain may be substantial in actual physiologic function.

7.2.2.3 Lung Walls. The elastic property like ligament is very important in lung tissue for obvious reasons. Also the surface properties of the tissue are inevitably related to the total function of the organ. Like skin, the lung is also exposed directly to the *ex vivo* environment; thus the protection against environment is quite important. The human lung has about 400 million alveoli supported mostly by elastic tissue component such as ligament. These membrane walls are bridged by fine collagen networks with numerous capillaries for efficient gas exchange.

A typical stress–strain curve for lung tissue is shown in Figure 7-16. As can be seen, the tissue can elongate 100 percent with small force, thus minimizing expansion energy of the lung. In real physiological conditions the volume–pressure relationship (Fig. 7-17) is more relevant. The pulmonary pressure changes of the lung can be expressed with a simple equation,

$$\Delta p = A(\Delta V) + B(\dot{V}) + C(\ddot{V}) \qquad (7\text{-}9)$$

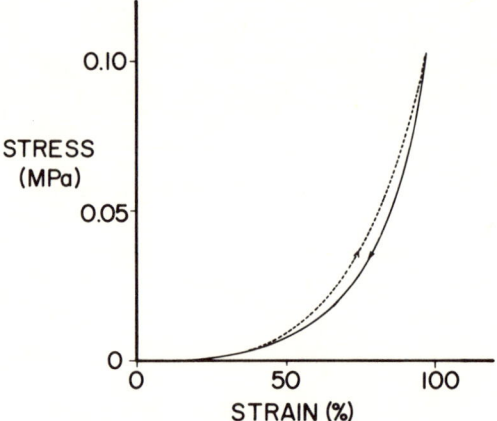

Figure 7-16. Stress–strain behavior of the alveolar wall of cat's lung. (Redrawn from H. Fukaya, C. J. Martin, A. C. Young, and S. Katsura, *J. Appl. Physiol.*, 25 (6), 689, 1968, by permission from the American Physiological Society.)

where A, B, and C are constant, V is volume change, $\dot{V}$ is velocity, and $\ddot{V}$ is acceleration of the air flow. For slow respiration the second and third terms become negligible, thus

$$\Delta p \simeq A(\Delta V) \qquad (7\text{-}10)$$

which indicates that the pressure change is in proportion to the volume change, and the constant (A) indicates the stiffness of the wall (Fig. 7-17).

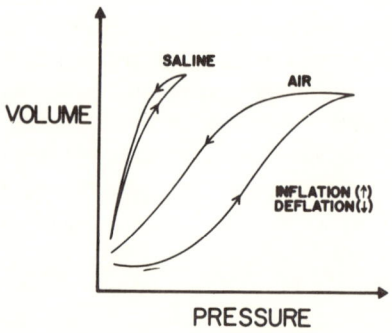

Figure 7-17. Volume–pressure relationship of excised lung filled with saline and air. Note the difference in the amount of volume. (Redrawn from G. C. Lee and F. G. Hoppin, Jr., in *Biomechanics,* chapter 14, ed. Y. C. Fung, N. Perrone, and M. Anliker, Prentice-Hall, Englewood Cliffs, N.J., 1972, by permission from the publisher.)

BIOLOGICAL MATERIALS

PROBLEMS

7-1. A sample of bone (0.50-cm diameter) broke with a load of 2600 N. Its final diameter is 0.49 cm.

a. What is its true breaking strength?
b. What is its engineering breaking strength?
c. What is the fracture strain?

Answers
Assuming that volume change is negligible:

a. $\quad \sigma = \dfrac{2600 \text{ N}}{\pi(0.245)^2 \times 10^{-4} \text{ m}^2} = \underline{137.9 \text{ MPa}}$

b. $\quad \sigma = \dfrac{2600 \text{ N}}{\pi(0.25)^2 \times 10^{-4} \text{ m}^2} = \underline{132.5 \text{ MPa}}$

c. $\quad \epsilon = \dfrac{\sigma}{E} = \dfrac{132.5 \text{ MPa}}{18 \text{ GPa}} = \underline{7.3 \times 10^{-3} \text{ (or 0.73\%)}}$

7-2. To test an artery, a bioengineer cut an artery into a 2-mm-wide sample similar to that of a rubber band. He tested the sample between two rollers in a saline solution of 37°C.

a. From the stress–strain curve determine the modulus of elasticity.
b. Explain some advantages and disadvantages of this type of test over the completely cut-out samples.

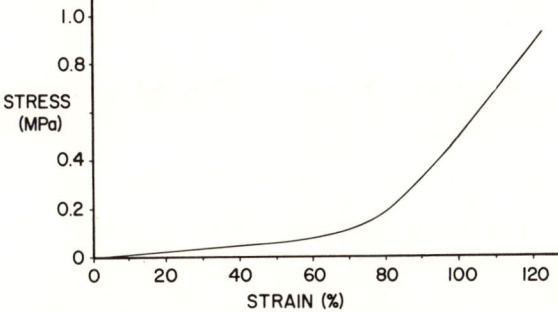

Engineering stress versus strain of human aorta.

Answers

a.
$$E_{\text{initial}} = \frac{0.07 \text{ MPa}}{0.55} = \underline{0.127 \text{ MPa}}$$

$$E_{\text{final}} = \frac{1 - 0 \text{ MPa}}{1.2 - 0.74} \quad \underline{2.17 \text{ MPa}}$$

b. The closer the tissue sample is tested to its natural state, the better and easier to interpret and use the data. Therefore, the rubber-band type of aortic specimen is better for the test. There is another problem of gripping the sample during testing. The stress concentration and the viscoelastic nature of the tissues make gripping a major problem. The rubber-band type of specimen is better for the test although it adds some other problems.

7-3. The bioengineer in problem 7-2 is trying to understand the contribution of smooth muscle toward mechanical properties of the artery and its relation with other tissue components such as collagen and elastin. The bioengineer chose a stress-relaxation test and artificial stimulation of the artery by using a drug (norepinephrine). The drug was introduced into the test chamber and the following figure was obtained:

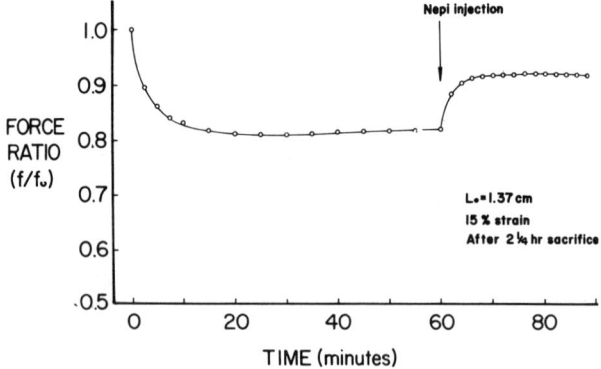

Stress-relaxation test on fresh canine artery. (Courtesy of A. S. Hoffman and J. B. Park.)

a. Assuming the tissue is kept alive at all times, calculate the contribution of the smooth muscle toward the total stress at the given strain (in this case 15%).
b. How fast is the drug (2 mg/ml of solution) acting on the aorta?
c. How can you keep the tissue alive?

Answers
a. Because the stress relaxed to 80% and returned to 90% by the stimulation of the muscle, the net contribution at this strain level is

BIOLOGICAL MATERIALS

$$\frac{0.9 - 0.8}{1 - 0.8} = 0.5 \text{ (or 50\%)}$$

b. It takes about 8 min to have a complete activation of the muscle and notice the continued contraction of the muscle.

c. The aorta can be kept alive for a day by using Krebs–Ringer solution with percolation of O_2—CO_2 gas.

7-4. In the same test as problem 7-3, the bioengineer used collagenase and repeated the same experiment. However, the collagenase took time to remove collagen in this test solution and therefore a control experiment had to be run, as shown in the following figures. (Assume that the collagenase does not affect the smooth muscle.)

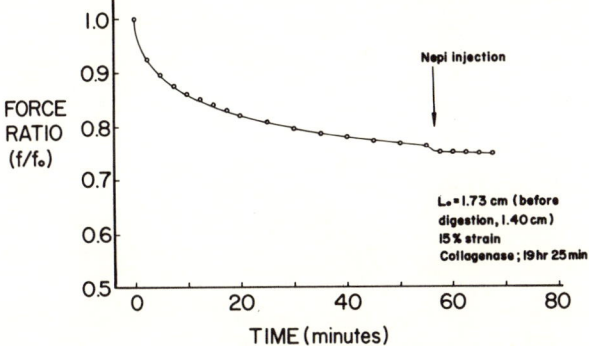

Stress-relaxation test on collagen-digested artery by collagenase.

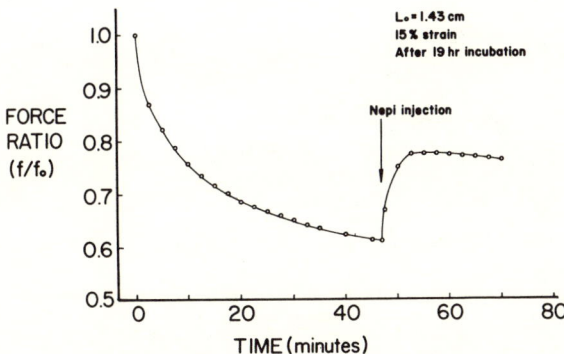

Stress-relaxation test on the specimen which underwent the same treatment as the collagenased one without removal of collagen.

a. Calculate the contribution of smooth muscle to the mechanical properties in the control sample.
b. Why does lesser relaxation occur for the collagenased sample?
c. Calculate the percent strain changes before and after collagenase. Does this affect the interpretation of the results?
d. What conclusion can you reach in this experiment?

Answers

a. $(0.78 - 0.61)/1 - 0.61 = 0.44$ (44%) (slight decrease of the muscle tone compared to the shorter period of storage from problem 7-3).
b. Collagen is a viscous substance that relaxes the stress; if this component is removed from the aorta, then the stress–relaxation should be minimal.
c. $(1.73 - 1.4)/1.4 = 0.24$ (or 24%). The length increased by 24% after collagenase treatment. However, since our concern is only with force ratios, this result does not affect our interpretation.
d. There is a definite link between the collagen and the smooth muscle in the aortic wall.

7-5. The viscoelastic properties of compact bone has been described by using the three-element model shown in the figure.

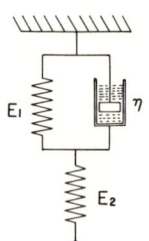

a. Derive differential equations for the system.
b. Solve the equation for ϵ and σ.
c. What are the shortcomings of the model? Explain them clearly.

Answers
a. *Differential equations:*
The total strain (ϵ) and stress (σ) can be written,

$$\epsilon = \epsilon_1 + \epsilon_2, \qquad \sigma = \sigma_1 = \sigma_2 \tag{1}$$

where subscripts 1 and 2 indicate element 1 (Voigt model with η and E_1) and element 2 (a spring, E_2).
Element 1 (Voigt model)
From equation (2-19),

$$\sigma_1 = E_1 \epsilon_1 + \eta \frac{d\epsilon_1}{dt} \tag{2}$$

Element 2

$$\sigma = E_2 \epsilon_2 \quad \text{hence} \quad \frac{d\epsilon_2}{dt} = \frac{1}{E_2} \frac{d\sigma}{dt} \tag{3}$$

BIOLOGICAL MATERIALS

From equation (2)

$$\frac{d\epsilon_1}{dt} = \frac{\sigma}{\eta} - \frac{E_1\epsilon_1}{\eta} \qquad (4)$$

Also from equation (1)

$$\frac{d\epsilon}{dt} = \frac{d\epsilon_1}{dt} + \frac{d\epsilon_2}{dt} \qquad (5)$$

(3), (4) → (5)

$$\frac{d\epsilon}{dt} = \frac{\sigma}{\eta} - \frac{E_1\epsilon_1}{\eta} + \frac{1}{E_2}\frac{d\sigma}{dt} \qquad (6)$$

Since $\epsilon_1 = \epsilon - \epsilon_2 = \epsilon - \sigma/E_2$

$$\frac{d\epsilon}{dt} = \frac{\sigma}{\eta} - \frac{E_1}{\eta}\left(\epsilon - \frac{\sigma}{E_2}\right) + \frac{1}{E_2}\frac{d\sigma}{dt} \qquad (7)$$

Rearranging (7)

$$\frac{d\sigma}{dt} + \frac{(E_1 + E_2)}{\eta}\sigma = \frac{E_1 E_2}{\eta}\epsilon + E_2\frac{d\epsilon}{dt} \qquad (8)$$

b. Solving the differential equation (8) according to the testing (boundary) conditions:

(1) Static testing.

$$\frac{d\epsilon}{dt}, \frac{d\sigma}{dt} \text{ are zero}$$

$$\frac{E_1 + E_2}{\eta}\sigma = \frac{E_1 E_2}{\eta}\epsilon$$

Therefore

$$\sigma = \frac{E_1 E_2}{E_1 + E_2}\epsilon$$

$$\underline{\sigma = E_s\epsilon} \qquad (9)$$

Equation (9) represents purely elastic behavior and E_s is called <u>static modulus of elasticity</u>.

(2) Creep testing. Instantaneously stretch the sample and hold the sample with a constant load (or weight). The instantaneous elastic strain is solely due to the element 2; thus $\epsilon_0 = \sigma/E_2$ and if we use this as our initial condition, then equation (8) can be written

$$E_2 \frac{d\epsilon}{dt} + \frac{E_1 E_2}{\eta} \epsilon = \frac{E_1 + E_2}{\eta} \sigma$$

Therefore

$$\frac{d\epsilon}{dt} + \frac{E_1}{\eta} \epsilon = \frac{E_1 + E_2}{\eta E_2} \sigma \qquad (10)$$

Thus the homogeneous solution is

$$\frac{d\epsilon}{dt} = -\frac{E_1}{\eta} \epsilon, \quad \frac{d\epsilon}{\epsilon} = -\frac{E_1}{\eta} dt$$

Integrating

$$\ln \epsilon = -\frac{E_1}{\eta} t + \text{constant}$$
$$\epsilon = A e^{-(E_1/\eta)t} \qquad (11)$$

for a particular solution; since the sample is held constant, stress and $d\epsilon/dt \approx 0$; thus equation (10) becomes

$$\frac{E_1}{\eta} \epsilon = \frac{E_1 + E_2}{\eta E_2} \sigma$$

Hence

$$\epsilon = \frac{E_1 + E_2}{E_1 E_2} \sigma = \frac{\sigma}{E_s} \qquad (12)$$

Combining (11) and (12)

$$\epsilon = \frac{\sigma}{E_s} + A e^{-(E_1/\eta)t}$$

At $t = 0$, $\epsilon = \epsilon_0 = \sigma/E_2$, hence $\sigma/E_2 = \sigma/E_s + A$; therefore

$$A = \sigma \left(\frac{1}{E_2} - \frac{1}{E_s} \right)$$

Therefore

$$\epsilon = \frac{\sigma}{E_s} + \sigma \left(\frac{1}{E_2} - \frac{1}{E_s} \right) e^{-(E_1/\eta)t} \qquad (13)$$

At $t = \infty$, $\epsilon_\infty = \sigma/E_s$; thus

$$\frac{\epsilon_\infty}{\epsilon_0} = \frac{\sigma/E_s}{\sigma/E_2} = \frac{E_2}{E_s} = \frac{E_2}{E_1 E_2/(E_1 + E_2)} = \frac{E_1 + E_2}{E_1} = 1 + \frac{E_2}{E_1} > 1$$

This indicates that the $\epsilon_\infty/\epsilon_0$ is dictated by the relative modulus of elements 1 and 2 for a given constant stress and is always greater than 1; i.e., the length increases with time.
(3) Stress relaxation testing. The sample is stretched instantaneously and held at constant length while monitoring the stress changes. The instantaneous stress on the body will be $\sigma_0 = E_2\epsilon$, because the viscous element cannot respond immediately. From equation (8)

$$\frac{d\sigma}{dt} + \frac{E_1 + E_2}{\eta} \sigma = \frac{E_1 E_2}{\eta} \epsilon \text{ since } \frac{d\epsilon}{dt} = 0 \tag{14}$$

Homogeneous solution is

$$\sigma = Be^{-[(E_1+E_2)/\eta]t} \quad (B = \text{constant}) \tag{15}$$

The particular solution is

$$\sigma = \frac{E_1 E_2}{E_1 + E_2} \epsilon = E_s \epsilon \tag{16}$$

Combining (15) and (16)

$$\sigma = E_s + Be^{-[(E_1+E_2)/\eta]t}$$

where $t = 0$, $\sigma = \sigma_0 = E_2\epsilon$; therefore, $B = (E_2 - E_s)\epsilon$ and

$$\sigma = E_s\epsilon + (E_2 - E_s)\epsilon e^{-[(E_1+E_2)/\eta]t} \tag{17}$$

At $t = \infty$, $\sigma_\infty = E_s\epsilon$; therefore

$$\frac{\sigma_\infty}{\sigma_0} = \frac{E_s\epsilon}{E_2\epsilon} = \frac{E_s}{E_2} = \frac{E_1 E_2/(E_1 + E_2)}{E_2} = \frac{E_1}{E_1 + E_2} < 1$$

The stress always decreases with time (i.e., stress relaxation).
(4) Constant strain rate test $\epsilon = Kt$ or $d\epsilon/dt = K$ (constant). From equation (8),

$$\frac{d\sigma}{dt} + \frac{E_1 + E_2}{\eta} \sigma = \frac{E_1 E_2}{\eta} Kt + E_2 K = E_2 K \left(1 + \frac{E_1 t}{\eta}\right) \tag{18}$$

From mathematical tables (*Standard Mathematical Tables*, 14th ed., p. 372, CRC Press, Cleveland, Ohio), $(D - a)y = P(x)$,

$$Y_P = -\frac{1}{a}\left[P(x) + \frac{P'(x)}{a} + \frac{P''(x)}{a^2} + \cdots + \frac{P^n(x)}{a^n}\right]$$

where $D = d/dt$, $y = \sigma$, $x = t$, $a = -(E_1 + E_2)/\eta$,

$$P(x) = E_2 K \left(1 + \frac{E_1 t}{\eta}\right), \qquad P'(x) = \frac{E_2 E_1 K}{\eta}$$
$$P''(x) = 0, \qquad P'''(x) = 0$$

Therefore

$$\begin{aligned}
\sigma_P &= +\frac{\eta}{E_1 + E_2}\left[E_2 K\left(1 + \frac{E_1 t}{\eta}\right) - \frac{\eta}{E_1 + E_2}\frac{E_2 E_1 K}{\eta}\right]\\
&= \frac{\eta}{E_1 + E_2}\left(E_2 K + \frac{E_1 E_2 K t}{\eta} - \frac{E_1 E_2 K}{E_1 + E_2}\right)\\
&= \frac{\eta E_2 K}{E_1 + E_2} + \frac{E_1 E_2 K t}{E_1 + E_2} - \frac{\eta E_1 E_2 K}{(E_1 + E_2)^2}\\
&= \eta \frac{E_s}{E_1} K + E_s K t - \eta \frac{E_s}{E_1 + E_2} K\\
&= \eta E_s K \left(\frac{1}{E_1} - \frac{1}{E_1 + E_2}\right) + E_s K t\\
&= \eta E_s K \frac{E_1 + E_2 - E_1}{E_1(E_1 + E_2)} + E_s K t\\
&= \eta E_s K \frac{E_2}{E_1(E_1 + E_2)} + E_s K t\\
&= K\eta \frac{E_1 E_2}{(E_1 + E_2)} \frac{E_2}{E_1(E_1 + E_2)} + E_s K t\\
&= K\eta \frac{E_2^2}{(E_1 + E_2)^2} + E_s K t
\end{aligned} \tag{19}$$

The homogeneous solution is

$$\sigma = Ce^{-[(E_1 + E_2)/\eta]t} \tag{20}$$

Combining equations (19) and (20),

$$\sigma = Ce^{-[(E_1 + E_2)/\eta]t} + K\eta \frac{E_2^2}{(E_1 + E_2)^2} + E_s K t$$

At $t = 0$, $\sigma = 0$ and

$$0 = C + K\eta \frac{E_2^2}{(E_1 + E_2)^2} + E_s K t$$

Therefore,

$$C = -K\eta \frac{E_2^2}{(E_1 + E_2)^2}$$

Final solution:

$$\begin{aligned}\sigma &= E_s K t + \frac{K\eta E_2^2}{(E_1 + E_2)^2} - \frac{K\eta E_2^2}{(E_1 + E_2)^2} e^{-[(E_1+E_2)/\eta]t} \\ &= E_s K t + \frac{K\eta E_2^2}{(E_1 + E_2)^2}(1 - e^{-[(E_1+E_2)/\eta]t})\end{aligned} \quad (21)$$

or since $t = \epsilon/K$,

$$\sigma = E_s \epsilon + \frac{K\eta E_2^2}{(E_1 + E_2)^2}(1 - e^{-[(E_1+E_2)/\eta](\epsilon/K)}) \quad (22)$$

If $K \approx 0$, static testing will have $\sigma = E_s \epsilon$; and if $K \approx \infty$, then $\sigma = E_2 \epsilon$. Since $E_2 > E_s$, the model fits the observation that increasing strain rate increases the modulus.

c. Shortcomings of the model

1. This equation describes only one-dimensional behavior.
2. Assumptions are not realistic; i.e., the bone is highly anisotropic and inhomogeneous (two or more phases present).
3. It cannot predict the fracture point.

FURTHER READING

R. Barker, *Organic Chemistry of Biological Compounds*, chapters 4 and 5, Prentice-Hall, Englewood Cliffs, N.J., 1971.
J. Black, "Dead or Alive: The Problem of *in Vitro* Tissue Mechanics," *J. Biomed. Mate. Res.* 10, 377, 1976.
M. Chvapil, *Physiology of Connective Tissue*, chapter 2, Butterworths, London, 1967.
H. R. Elden (ed.), *Biophysical Properties of the Skin*, J. Wiley and Sons, New York, 1971.
H. Fleisch, H. J. J. Blackwood, and M. Owen (eds.), *Calcified Tissue*, Springer-Verlag, New York, 1966.
Y. C. Fung, N. Perrone, and M. Anliker (eds.), *Biomechanics, Its Foundations and Objectives*, Prentice-Hall, Englewood Cliffs, N.J., 1972.
K. H. Gustavson, *The Chemistry and Reactivity of Collagen*, Academic Press, New York, 1956.

D. A. Hall, *The Chemistry of Connective Tissue,* Charles C Thomas, Springfield, Ill., 1961.

R. M. Kenedi (ed.). *Perspectives in Biomedical Engineering,* University Park Press, Baltimore, 1973.

H. Kraus, "On the Mechanical Properties and Behavior of Human Compact Bone," *Advan. Biomed. Eng. Med. Phys., 2,* 169, 1968.

J. B. Park, C. H. Daly, and A. S. Hoffman, "The Contribution of Collagen to the Mechanical Response of Canine Artery at Low Strains," *Frontiers of Matrix Biology,* vol. 3, ed. A. M. Robert and L. Robert, p. 218, Karger, Basel, Switzerland, 1976.

G. N. Ramachandran (ed.), *Treatise on Collagen,* volume 2A, chapter 6, "Mechanical Properties of Collagenous Tissues," by R. D. Harkness; volume 2B, chapter 3, "Organization and Structure of Bone," by M. J. Glimcher and S. M. Krane; volume 1, chapter 1, "Composition of Collagen and Allied Proteins," by J. E. Eastoe, Academic Press, New York, 1967 and 1968.

J. W. Remington (ed.), *Tissue Elasticity,* American Physiological Society, Washington, D.C., 1957.

A. Viidik, "Functional Properties of Collagenous Tissues," *Intern. Rev. Connect. Tissue Res., 6,* 127 (1973).

G. L. Wilkes, I. A. Brown, and R. H. Wildnauer, "The Biochemical Properties of Skin," *CRC Crit. Rev. Bioeng., 1*(4), 459, 1973.

H. Yamada, *Strength of Biological Materials,* Williams & Wilkins Co., Baltimore, 1970.

I. Zipkin (ed.), *Biological Mineralization,* J. Wiley and Sons, New York, 1973.

CHAPTER 8

TISSUE RESPONSE TO IMPLANTS

To implant a material the tissue must be injured first. The injured or diseased tissues should be removed to some extent in the process of implantation. The success of the entire operation depends on the kind and degree of tissue response to implants during the healing process. The tissue response to injury may vary widely according to the site, species, contamination, etc. However, the inflammation and the cellular response to the wound for both intentional and accidental injuries are the same regardless of the sites.

8.1. WOUND HEALING PROCESS

8.1.1. Inflammation

Whenever tissues are injured or destroyed the adjacent cells respond to repair them. An immediate response to any injury is the acute inflammatory reaction. Soon after constriction of capillaries (stopping blood leakage) dilatation occurs while there is a greatly increased activity in the endothelial cells lining the capillaries. The capillaries will be covered by adjacent leukocytes, erythrocytes, and platelets. Concurrently with vasodilatation, leakage of plasma from capillaries occurs. The leaked fluid combine with the migrating leukocytes and dead tissues constitute exudate. Once enough polymorphonuclear cells are accumulated by lysis, the exudate becomes pus. (Table 8-1 defines types of cells.) It is important to know that the pus can sometimes occur in nonbacterial inflammation.

Table 8-1. Definitions of Cells

Cells	Description
Chondroblast	An immature cell producing collagen (cartilage).
Endothelial cell	A cell lining the cavities of the heart and the blood and lymph vessels.
Erythrocyte	A formed element of blood containing hemoglobin (red blood cells).
Fibroblast	A common fixed cell of connective tissue that elaborates the precursors of the extracellular fibrous and amorphous components.
Giant cell	
Foreign body giant cell	A large cell derived from a macrophage in the presence of a foreign body.
Multinucleated giant cell	A large cell with many nuclei.
Granulocyte	Any blood cell containing specific granules; included are neutrophils, basophils, and eosinophils.
Leukocyte	A colorless blood corpuscle capable of ameboid movement, protects body from microorganisms and can be one of five types: lymphocyte, monocyte, neutrophil, eosinophil, and basophil.
Macrophage	Large phagocytic mononuclear cell. Free macrophage is an ameboid phagocyte and present at the site of inflammation.
Mesenchymal cell	Undifferentiated cell with similar roles as fibroblasts but often smaller; can develop into new cell types by certain stimuli.
Mononuclear cell	Any cell with one nucleus.
Osteoblast	An immature bone-producing cell.
Phagocyte	Any cell that destroys microorganisms or harmful cells.
Platelet	A small circular or oval disk ~ 3-μm diameter found in the blood; elaborates in coagulation of blood.

At the time of damage to the capillaries the local lymphatics are also damaged because they are more fragile than the capillaries. However, the leakage of fluids from capillaries will provide fibrinogen and other formed elements of the blood clotting system (see Chapter 10) that will quickly plug the damaged lymphatics thus localizing the inflammatory reaction.

All these reactions—vasodilatation of capillaries, leakage of fluid into the extravascular space, and plugging of lymphatics—provide the classic inflammatory signs: redness, swelling, and heat which can lead to local pain.

When the tissue injury is extensive or the wound contains either irritants or bacteria, the inflammation may lead to extensive tissue destruction. The tissue destruction is done by the collagenase, which is a proteolytic enzyme capable of digesting collagen. The collagenase is released from granulocytes which in turn are lysed by the lower pH at the site of wound. Local pH can be dropped to below 5.2 at the injured site from the normal values of 7.4–7.6. If there is no drainage for the necrotic debris, lysed granulocytes, formed blood elements, etc., then the site becomes a severely destructive inflammation resulting in a necrotic abscess.

If the severely destructive inflammation persists without healing, a chronic inflammatory process commences within three to five days. This is marked by the presence of mononuclear cells called macrophages, which can coalesce to form multinuclear giant cells. The macrophages are phagocytic and remove foreign materials or bacteria. Sometimes the mononuclear cells evolve into histiocytes that regenerate collagen. This regenerated collagen is used to unite a wound or wall off unremovable foreign materials by encapsulation.

In chronic inflammatory reaction lymphocytes occur as clumps or foci. These cells are a primary source of the immunogenic agents that become active if foreign proteins are not removed by the body's primary defense. An autoimmune reaction is suggested as a foreign body reaction to nonproteinous materials such as silica.

8.1.2. Cellular Response to Repair

Soon after injury the mesenchymal cells evolve into migratory fibroblasts that move into the injured site while the necrotic debris, blood clots, etc., are removed by the granulocytes and macrophages. The inflammatory exudate contains fibrinogen, which is converted into fibrin by enzymes released through blood and tissue cells (Chapter 10). The fibrin scaffolds the injured site. The migrating fibroblasts use the fibrin scaffold as a framework onto which the collagen is deposited. New capillaries are formed following the migration of fibroblasts and the fibrin scaffold is removed by the fibrinolytic enzymes activated by the endothelial cells. The endothelial cells together with the fibroblasts liberate collagenase, which limits the collagen content of the wound.

After two to four weeks of fibroblastic activities the wound undergoes remodeling by decreasing the glycoprotein and polysaccharide content of the scar tissue and lowering the number of synthesizing fibroblasts. A new balance of collagen synthesis and dissolution is reached and the maturation phase of the wound begins. The time required for the wound healing process varies with various tissues, although the basic steps described here can be applied in all connective tissue wound healing processes.

The healing of soft tissues, especially the skin wound, has been studied intensively because it is germane to all surgery. The degree of healing can be determined by histochemical or physical parameters. A combined method will give a better understanding of the overall healing

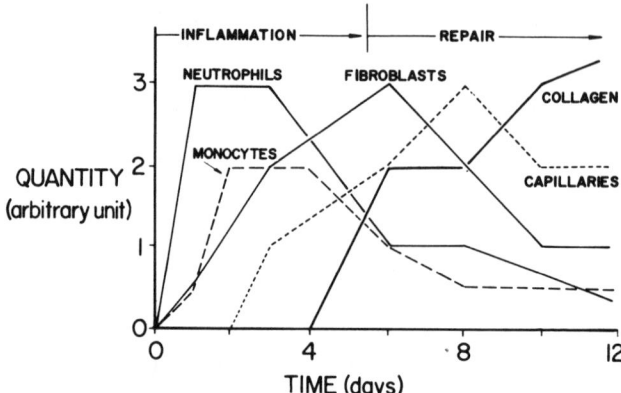

Figure 8-1. Soft tissue wound healing sequence. (Modified from R. Ross, *Sci. Amer.*, *220*, 40, 1969.)

process. Figure 8-1 shows a schematic diagram of sequential events in cellular responses of soft tissues. The wound strength is not proportional to the amount of collagen deposited in the injured site as shown in Figure 8-2. This indicates that there is a latent period for the collagen molecules (procollagen is deposited by fibroblasts) to polymerize. It may take additional time to align the fibers in the direction of stress and cross-link fibrils to increase the physical strength closer to a normal tissue. This collagen restructuring process requires more than six months to complete although the wound strength never reaches the original value. The wound strength can be affected by many variables, e.g., severe malnutrition, resulting in protein depletion, temperature, other wounds, and oxygen tension. Other factors such as drugs, hormones, irradiation, and electrical stimulation all affect the normal wound healing process.

The healing of bone fracture is regenerative rather than the simple repair seen in other tissues (except liver). However, the extent of regeneration is limited in humans. The cellular events following fracture of bone are illustrated in Figure 8-3. When a bone is fractured, many blood vessels (including the adjacent soft tissues) hemorrhage and form a blood clot around the fracture site. As in any wound repair, shortly after fracture the fibroblasts in the outer layer of periosteum and the osteogenic cells in the inner layer of periosteum migrate and proliferate toward the injured site. These cells lay down a fibrous collagen matrix called callus. Osteoblasts evolved from the osteogenic cells near the bone surface start to calcify the callus into trabeculae, forming a spongy bone. The osteogenic cells migrating farther away from an established blood supply become chondro-

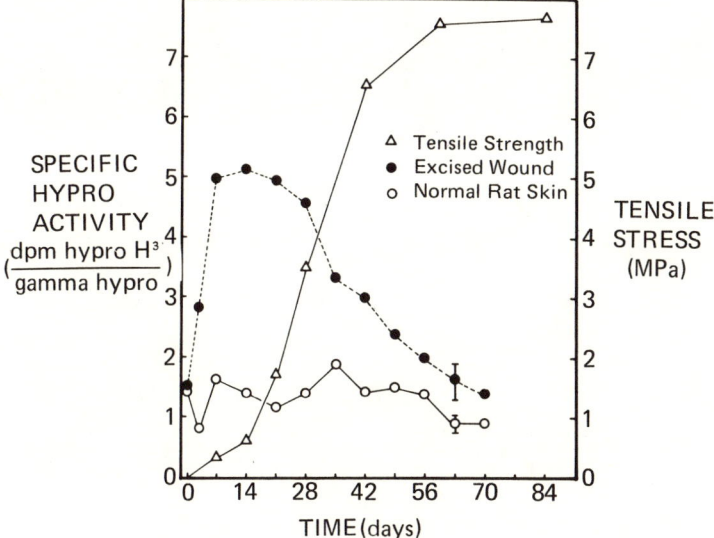

Figure 8-2. Tensile strength and rate of collagen synthesis of rat skin wounds. (Redrawn from E. E. Peacock, Jr., and W. Van Winkle, Jr., *Surgery and Biology of Wound Repair*, p. 140, W. B. Saunders Co., Philadelphia, 1970, by permission from the publisher.)

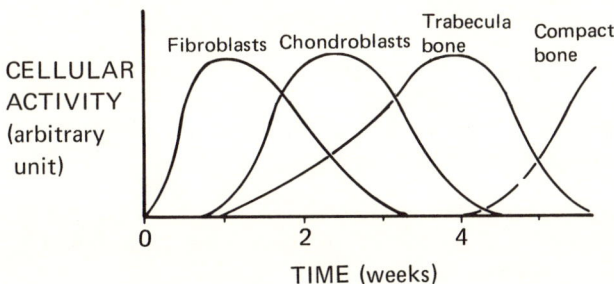

Figure 8-3. Sequence of events following bone fracture. (After L. L. Hench and E. C. Ethridge, *Advan. Biomed. Eng., 5,* 35, 1975, by permission from the publisher.)

blasts, which lay down cartilage. Thus after about two to four weeks the periosteal callus is made of three parts, as shown in Figure 8-4.

Simultaneous with the external callus formation a similar repair process occurs in the marrow cavity. Because blood is in abundant supply the cavity turns into callus rather fast and becomes fibrous or spongy bone.

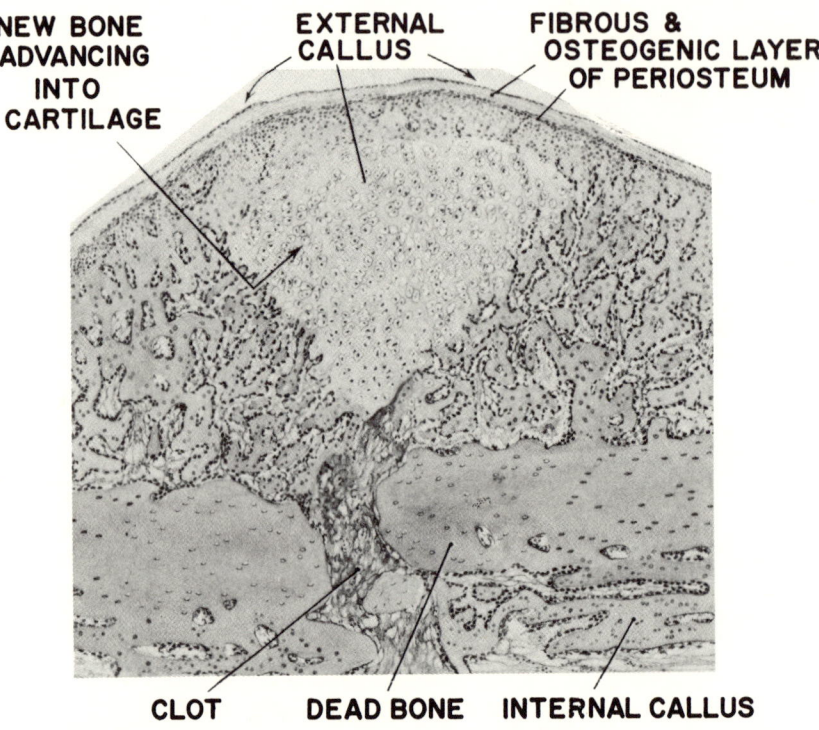

Figure 8-4. Part of a longitudinal section of fractured rib of a rabbit after two weeks, H & E stain. (Reproduced by permission from A. W. Ham and W. R. Harris, in *Biochemistry and Physiology of Bone,* ed. G. H. Bourne, Academic Press, New York, 1956.)

A new trabeculae develop in the fracture site by appositional growth and the spongy bone turns into compact bone. This maturation process begins after about four weeks.

Some other interesting observations have been made on the healing of bone fractures in relation to the synthesis of polysaccharide and collagen. It is believed that the amount of collagen and polysaccharides is closely related to the cellular events following fracture. When the amount of collagen starts to increase, it marks the onset of the remodeling process. This occurs after about one week. Another interesting observation is the electric potential (or biopotential) measured in the long bone before and after fracture, as shown in Figure 8-5. The large electronegativity in the vicinity of the fracture marks increased cellular activity in the tissues. Thus there is a maximum negative potential in the epiphysis in normal bone because this zone is more active (growth plate is in the epiphysis).

TISSUE RESPONSE TO IMPLANTS

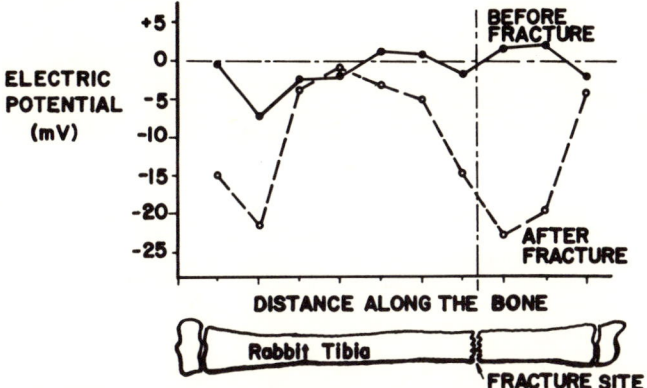

Figure 8-5. Skin surface potentials of rabbit limb before and after fracture. Note that the fracture site increases electronegative potential. (Redrawn from Z. B. Friedenberg and C. T. Brighton, *J. Bone Joint Surg.*, *48A*, 915, 1966, by permission from the publisher.)

Example 8-1

The healing process of skin wounds has been investigated many times since it is somewhat related to every type of surgery. In one study electrical stimulation was used to accelerate rabbit skin wound, as shown in the figure. The mean current flow was 21 μA and the mean current density was 8.4 μA/cm^2. After 7 days, the load to fracture on the control samples gave 797 g and the stimulated experimental side gave 1224 g on the average [J. J. Konikoff, *Ann. Biomed. Eng.*, *4*, 1(1976)].

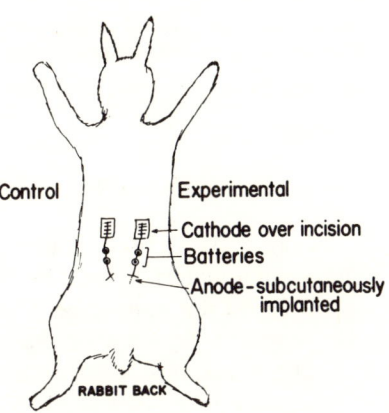

Schematic diagram of skin wound stimulation experiment.

a. Calculate the percent increase strength by stimulation.
b. The width of the testing sample was 1.6 cm. Assuming 1.8-mm thickness of skin, calculate the tensile stress for both control and experimental sample.
c. Compared with the strength of normal skin (about 8 MPa), what percentages of the control and experimental skin wound strengths were recovered?
d. Compare the results of (c) with the result of Figure 8-2.

Answers

a. $\dfrac{1224 - 797}{797} = \underline{0.536\ (53.6\%)}$

b. $\text{Stress} = \dfrac{797 \times 10^{-3} \times 9.8\text{N}}{1.8 \times 16 \times 10^{-6}\ \text{m}^2} = \underline{0.27\ \text{MPa}}$

$\text{Stress} = \dfrac{1{,}224 \times 10^{-3} \times 9.8\text{N}}{1.8 \times 16 \times 10^{-6}\ \text{m}^2} = \underline{0.42\ \text{MPa}}$

c. $\dfrac{0.27}{8} = \underline{0.034\ (\text{or } 3.4\%)}$, $\quad \dfrac{0.42}{8} = \underline{0.052\ (5.2\%)}$

d. About the same recovery.

8.2. BODY RESPONSE TO IMPLANTS

The body's response to implants varies widely according to the host site and species, the degree of trauma during implantation, and all the variables associated with normal wound healing. On the other hand, the chemical composition and micro- and macrostructure of the implants induce different body response. Body response has been studied in two different areas: local (cellular) and systemic, although a single implant should be tested for both aspects. In practice such studies are not done simultaneously except in a few cases such as bone cement.

8.2.1. Cellular Response to Implants

Generally the body reacts to the foreign materials to get rid of them. The foreign material could be extruded from the body if it can be moved or walled off if it cannot be moved. If the material is particulate or fluid, then it will be ingested by the giant cells (macrophages) and removed.

These responses are related to the healing process of the wound where the implant is added as an additional factor. A typical tissue response is that the polymorphonuclear leukocytes appear near the implant followed by the macrophages called foreign body giant cells. However, if the implant is chemically and physically inert to the tissue then the foreign body giant cells may not form. Instead, only a thin layer of collagenous tissue encapsulates the implant. If the implant is either a chemical or a physical irritant to the surrounding tissue then the inflammation occurs in the implant site. The inflammation (both acute and chronic) will delay the normal healing process, resulting in granular tissues. Some implants may cause necrosis of tissues by chemical, mechanical, and thermal trauma.

Table 8-2. Properties of Various Suture Materials *in Vivo*[a]

Material	Wound tensile strength	Suture tensile strength	Tissue reaction
Absorbable			
Plain catgut	Impaired	Zero by 3–6 days	Very severe
Chromic catgut	Impaired	Variable	Moderate (much less severe than plain catgut, but more than nonabsorbable materials)
Nonabsorbable			
Silk	No effect	Well maintained	Slight
Nylon multifilament	No effect	Very low at 6 months	Moderately severe and prolonged
Nylon monofilament	No effect	Well maintained	Slight
Polyethylene terephthalate	No effect	Well maintained	Very slight
PTFE (Teflon®)	No effect	Well maintained	Almost none

[a] After J. K. Newcombe, "Wound Healing," in *Scientific Basis of Surgery*, 2nd edition, ed. W. T. Irvin, p. 371, J & A Churchill, London, 1972, by permission from the publisher.

It is generally very difficult to assess tissue responses to various implants because of wide variations in experimental protocols. Table 8-2 describes the tissue reactions for various suture materials.

The degree of tissue response varies according to both the physical and chemical nature of the implants. Pure metals evoke a severe tissue reaction, which may be related to the high-energy state or large free energy of pure metals that tend to lower their free energy by oxidation or corrosion. In fact, the low tissue reaction exhibited by titanium and aluminum results from the tenacious oxide layer, which resists further diffusion of metal ions and gas (O_2) at the interface. In fact, this oxide layer makes such metals ceramiclike materials, which are very inert. Corrosion-resistant metal alloys such as cobalt-chromium and 316L stainless steel have similar effect on the tissue.

Most ceramic materials investigated for their tissue compatibility are such oxides as TiO_2, Al_2O_3, ZrO_2, and $BaTiO_3$ and multiphase ceramics of $CaO-Al_2O_3$, $CaO-ZrO_2$, and $CaO-TiO_2$. These materials showed minimal tissue reactions with only a thin layer of encapsulation, as shown in Figure 8-6. Similar reactions were seen for carbon implant.

Some glasses (e.g., 45 w/o SiO_2, 24.5 w/o Ca_2O, 24.5 w/o CaO, and

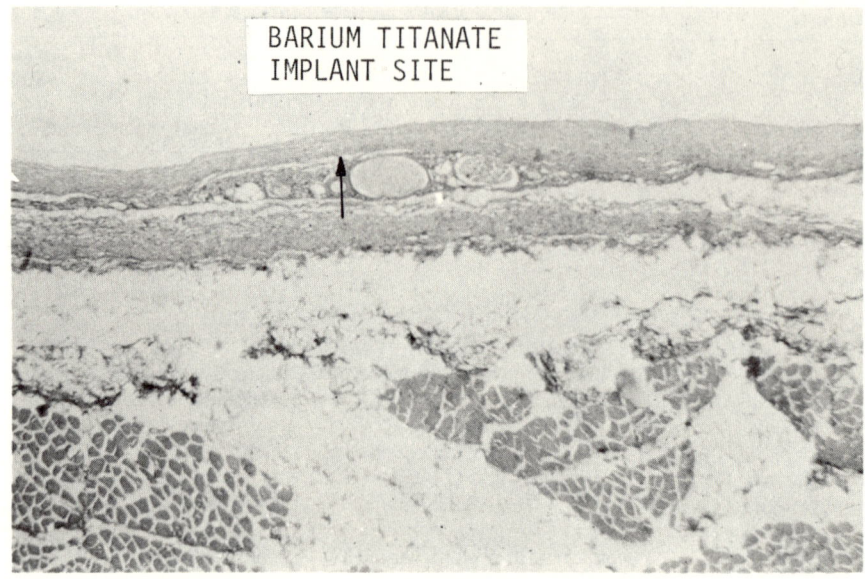

Figure 8-6. Optical micrograph of soft connective tissue adjacent to the BaTiO$_3$ implants after 20 weeks. [H & E stain, arrow (↑) indicates encapsulation of dense collagenous tissue, 400×.] (Courtesy of G. H. Kenner and J. B. Park.)

6.0 w/o P$_2$O$_5$) show a direct bonding between implant and bone by dissolution of the silica-rich gel film at the interface, as shown in Figure 8-7.

The polymers as such are quite inert in tissue if there are no additives such as antioxidant, fillers, antidiscoloring agents, or plasticizers. On the other hand, the monomers can evoke an adverse tissue reaction because they are reactive. Thus the degree of polymerization is somewhat related to the tissue reaction. Since a 100% polymerization is almost impossible to achieve, there is a range of different-size polymer molecules that can be leached out of the polymer. This leads to the fact that the particulate form of very inert polymeric materials can cause severe tissue reaction. This is amply demonstrated by polytetrafluoroethylene (Teflon®), which is quite inert in such bulk form as rods or woven fabric but very reactive in tissue when made into powder form. A schematic summary of tissue responses to implants is given in Figure 8-8.

There has been some concern about the possibility of tumor formation resulting from the wide range of materials used in implantation. Although many implant materials are carcinogenic in rats, there is no documented case of human tumors directly related to implants. It may be too

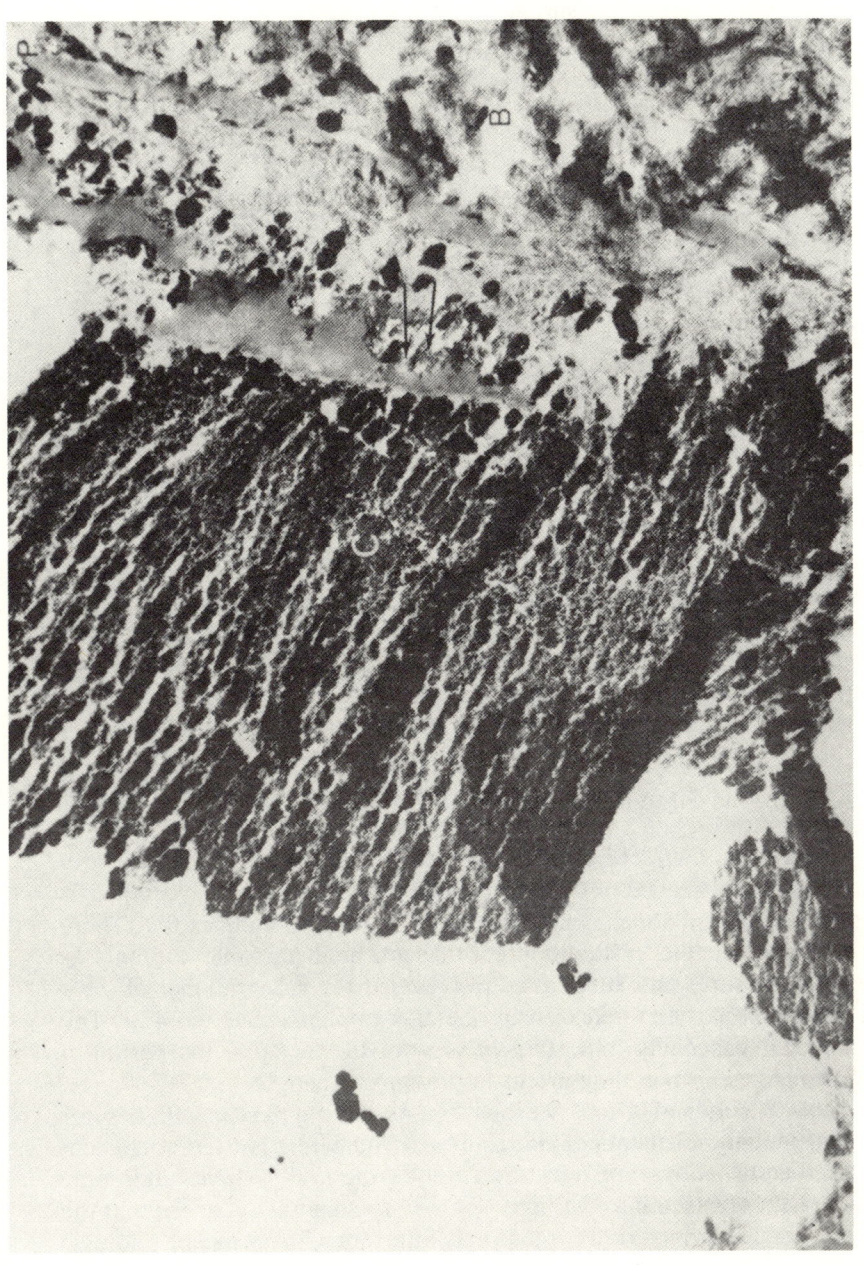

Figure 8-7. Electron micrograph of the interface between 45S5 bioglass–ceramic (C) and bone (B). The arrows indicate region of gel formation, undecalcified. (10,000×). (From L. L. Hench and H. A. Paschall, *J. Biomed. Mater. Res. Symp.*, No. 5, 49, 1974, courtesy of J. Wiley and Sons, New York.)

```
┌─────────────────────────────────────────────────────────────────┐
│                      │ IMPLANT: TISSUE │                        │
│                                                                 │
│                    MINIMAL RESPONSE                             │
│                 Thin Layer of Fibrous Tissue                    │
│              Silicone rubber, polyolefins, PTFE (Teflon),       │
│              PMMA, most ceramics, Ti & Co–Cr alloys             │
│                                                                 │
│  CHEMICALLY INDUCED RESPONSE      PHYSICALLY INDUCED RESPONSE   │
│  Acute, Mild Inflammatory Response   Inflammatory Response to Particulates │
│    Absorbable sutures, Some thermo-     PTFE, PMMA, nylon, metals │
│    setting resins                                               │
│  Chronic, Severe Inflammatory Response  Tissue Growth into Porous Materials │
│    Degradable materials, thermoplastics  Polymers, ceramics, metals │
│    with toxic additives, corrosion metal                        │
│    particles                                                    │
│                    NECROTIC RESPONSE                            │
│                 Layer of Necrotic Debris                        │
│                 Bone cement, surgical adhesives                 │
└─────────────────────────────────────────────────────────────────┘
```

Figure 8-8. A brief summary of tissue response to implants. (Modified from D. F. Williams and R. Roaf, *Implants in Surgery*, p. 233, W. B. Saunders Co., London, 1973, by permission from the publisher.)

premature to pass the final judgment because the latent time for human tumor formation may be longer than 20 to 30 years. However, the number of implants placed in the body with no report of direct evidence yet tends to support the observation that the carcinogenesis is species-specific and that no tumors will be formed in humans by implants.

8.2.2. Systemic Effects by Implants

The systemic effect by implants is well documented in hip joint replacement surgery. The polymethylmethacrylate bone cement applied in the femoral shaft in dough state lowers blood pressure significantly. There is concern about the systemic effect of such biodegradable implants as absorbable sutures and surgical adhesives and the large number of wear and corrosion particles released by metallic implants (Section 4.3). The latter fact is especially important in view of the fact that the period of implant is longer when they are used in younger people.

Table 8-3 indicates that various organs have different affinities for different metallic elements. This result also indicates that the corrosion-resistant metal alloys are not completely structurally stable and some elements are released into the body. This causes another concern that the elevated ion levels in various organs may interfere with normal physiologic activities. The divalent metal ions may also inhibit various enzyme activities.

Table 8-3. Concentrations of Metals in Tissue and Organs of the Rabbit after Implantation of Various Metals

Vitallium® (61.9% Co, 28 to 34% Cr, 4.73% Mo, 1.52% Ni, 0.61% Fe)
(2 or 3 specimens per experiment)

	Surrounding muscle		Liver		Kidney		Spleen		Lung		Control muscle	
	6 wk	16 wk	6 wk	16 wk	6 wk	16 wk	6 wk	16 wk	6 wk	16 wk	6 wk	16 wk
Chromium	30,25,0	0,0,15	0.5,5	5,5,10	0.5,5	5,5,10	0,0.5	5,5,10	0,0,10	5,5,10	0.5,5	0,10,0
Cobalt	25,45	0,30,0	5,10	5,5,10	0,0,105	5,10,70	5,5,20	5,5,300	0,0,300	10,5,10	0,0,50	0,10,0
Nickel	20,35	10	5,10	5,5,80	5,110	5,5,30	5,205	5,5,1000	5,40	5,0,45	5,5	5
Titanium	5,5,10	0,10,0	0,0.5	0,0,10	0,0,10	0,0,10	0,0.5	0,0,15	0,0,5	0,0,15	0,0,10	0,0,5
Molybdenum	0.5,5	0.5,0	0,15,80	10,10,70	0,20,50	20,20,75	0,0,10	0,0.5	0,0,10	5,5,5	0,0,0	0,0,0
Iron	0,40,90	0,80,0	0,200,600	100,160,80	0,90,180	110,120,90	0,250,270	300,290,110	0,90,420	110,100,50	0,10,30	0,20,0
316 Stainless steel (17.8% Cr, 13.4% Ni, 2.3% Mo, 0.23% Cu, 66.27% Fe)												
Chromium	145,65	115,295	5,10	5,5	5,5	5,5	5,5	5,20	5,5	20,10	0,20	5,5
Cobalt	5,5	0,10	5,20	0,15	5,10	5,115	5,5	15,600	5,5	0,250	0,0	0,90
Nickel	15,50	200,70	20,20	0,95	0,10	10,10	5,10	1000,65	20,220	20,45	10,0	5,5
Titanium	20,10	25,10	0,10	5,5	0,5	5,65	0,5	10,50	0,10	5,5	5,10	10,10
Molybdenum	5,10	10,35	10,75	50,65	10,85	75,80	5,10	15,150	0,10	5,5	0,0	0,0
Iron	80,70	90,190	220,590	500,520	110,210	180,200	180,420	580,220	120,220	150,300	30,10	15,10

[a] Taken from A. B. Ferguson, Y. Akahoshi, P. G. Laing, and E. S. Hodge, *J. Bone Joint Surg.*, **44A**, 323, 1962, by permission from the publisher.
[b] Figures are in ppm dry ash, given to nearest 5 ppm.

As mentioned before, the polymeric materials contain additives that cause cellular as well as systemic reactions to a greater degree than the polymer itself.

Even a well-accepted polymer dimethylsiloxane (Silastic®, Dow Corning Co.) contains a filler, silica powder, to increase the mechanical strength. Although the silica power itself is reactive when implanted in concentrated area, there seems no problem with this additive. However, it is not certain whether there will be a late complication if a large amount of the silica is released into the tissue and retained in various organs.

Example 8-2

A biodegradable suture will have a strength one-tenth of the original after 3 weeks of implantation. The strength of the implanted suture decreases according to

$$\sigma = \sigma_o + b \ln t$$

where σ_o = 10 MPa, b = −2 MPa/week, and t = time in week. Determine how long it was implanted.

Answer

$$\sigma = \sigma_o + b \ln t$$

$$\ln t = \frac{9}{2}$$

$$t = \exp(4.5) = \underline{42.65 \text{ weeks}} \text{ (or 297 days)}$$

PROBLEMS

8-1. Some materials do not normally induce tissue reaction as a bulk. However, when implanted in powder form, they become nonbiocompatible. Explain why.

Answer
The increased surface area when made into powder form increases the surface energy a great deal—to which the tissue has to counteract. This will sharply increase cellular activities and the body tries to reject individual particles.

8-2. Sometimes the degree of tissue reaction toward an implant is represented by the thickness of the collagenous fibrous capsule (e.g., Figure 8-6). State what fallacy these experimental results may have for deciding biocompatibility. Do you think the thicker the capsulation the better or worse as an implant?

Answer
The thickness of the collagenous tissue depends largely on the weight of the implant and its movement during implantation, not on its intrinsic properties. Therefore, the thickness

measurement is not a suitable method. Generally, the thinner the encapsulation of the implant, the better.

8-3. Calculate the average concentration of metals when Vitallium® is implanted (Table 8-2) in surrounding muscle and kidney at 6 and 16 weeks. Normalize the averages according to their weight percentage in the alloy.

Answer

	Muscle				Kidney			
	6 weeks		16 weeks		6 weeks		16 weeks	
Element	Average	Normalized	Average	Normalized	Average	Normalized	Average	Normalized
Cr	38	122.6	5	16	3.3	10.6	6.3	20
Co	30	48.5	7.5	12	35	56.5	28.3	45.7
Ni	28	1842	7.5	493	13.3	875	13.3	875
Mo	3.3	70	31.7	670	23.3	492	38.3	810
Fe	43	7049	267	43,770	90	14,754	106.7	17,491

8-4. Explain why metals are generally less biocompatible than ceramics or polymers. What can you do to improve this disadvantage of metals as implant material?

Answer
The higher-energy state of metals tends to make them more reactive with tissue than ceramics, which are already oxidized, and polymers, which have giant molecules (chains).

8-5. The temperature changes resulting from the heat of polymerization of bone cement (polymethylmethacrylate–polystyrene copolymer powder plus methylmethacrylate monomer liquid) was monitored at the interface between bone and cement (placed in the canine femur as a 9-mm-diameter plug), as shown in the following figure.

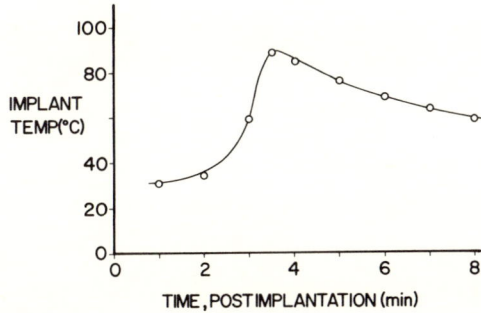

Temperature rise at the cement–bone interface in canine femur. The cement was mixed outside the body for 2 minutes before insertion. (Redrawn from C. A. Homsy, in *Biomaterials*, ed. A. L. Bement, Jr., p. 138, University of Washington Press, Seattle, 1971.)

a. What will happen to the adjacent tissues by the heat generated by the polymerization?
b. Would the temperature rise or decrease by putting a metal cylinder in the middle similar to the situation of femoral hip replacement?
c. What problems will rise if the cement shrinks when it reaches ambient temperature?

Answers
a. Because the maximum temperature reached is well over 55°C, the temperature at which proteins will begin to denature, the adjacent tissues will be damaged.
b. The temperature will decrease because the metal acts as a heat sink.
c. A gap will be developed between the bone and the cement (see Example 12-1).

FURTHER READING

C. O. Bechtol, A. B. Ferguson, and P. G. Laing, *Metals and Engineering in Bone and Joint Surgery,* Ballière, Tindall, and Cox, London, 1959.

G. H. Bourne (ed.), *The Biochemistry and Physiology of Bone,* vol. III, chapter 10, 2nd edition, Academic Press, New York, 1971.

J. Charnley, *Acrylic Cement in Orthopaedic Surgery,* Churchill Livingstone, Edinburgh, 1970.

T. Gillman, "On Some Aspects of Collagen Formation in Localized Repair and in Diffuse Fibrotic Reactions to Injury," in *Treatise on Collagen,* ed. B. S. Gould, vol. 2B, chapter 4, Academic Press, New York, 1968.

L. L. Hench and E. C. Ethridge, "Biomaterials—The Interfacial Problem," *Adv. Biomed. Eng., 5,* 35, 1975.

S. F. Hulbert, S. N. Levine, and D. D. Moyle, *Prosthesis and Tissue: The Interface Problem,* J. Wiley and Sons, New York, 1974.

S. N. Levine (ed.), "Materials in Biomedical Engineering," *Ann. N.Y. Acad. Sci., 146,* 1968.

H. I. Maibach and D. T. Rovee (eds.), *Epidermal Wound Healing,* Year Book Medical Publishers, Chicago, 1972.

E. E. Peacock, Jr., and W. Van Winkle, Jr., *Surgery and Biology of Wound Repair,* chapters 1, 5, 6, and 11, W. B. Saunders Co., Philadelphia, 1970.

R. Ross, "Wound Healing," *Sci. Am., 220,* 40, 1969.

CHAPTER 9

SOFT TISSUE REPLACEMENT I: SUTURES, SKIN, AND MAXILLOFACIAL IMPLANTS

The success of soft tissue implants has primarily been due to the development of synthetic polymers. This is mainly because the polymers can be tailored to match the physical and chemical properties of soft tissues. Polymers can be made into various physical forms such as liquid for filling spaces, fibers for suture materials, films for catheter balloons, knitted fabrics for blood vessel prostheses, and solid forms for cosmetic and weight-bearing applications.

Obviously, different applications demand different material properties. The following list outlines minimal requirements for all soft tissue replacements.

1. The implants should closely approximate physical properties, especially in flexibility and texture.
2. The implants should not deteriorate.
3. The implants should not cause severe tissue reaction. Although some say that a minor tissue reaction may be beneficial for faster wound healing as in the case of nylon sutures, it is generally accepted that minimum tissue reaction is most desirable.
4. The implants should not induce fibrous tissue encapsulation or ingrowth. Such ingrowth causes the loss of the originally intended function of the implant by the collagenous tissues. The

marble breast of the early mammary implant was a result of this effect. An exception to this rule is the blood vessel prosthesis, where success depends on the formation of a pseudoendothelial layer to prevent blood clotting or emboli formation.

5. The implants should be noncarcinogenic, nonallergic, and nonimmunogenic.

Other important factors are sterilizability, feasibility of mass production, cost, fatigue life, aesthetic quality, etc. This chapter deals mainly with some biomaterials used for soft tissue replacements such as sutures, artificial skins, and percutaneous devices.

9.1. SUTURES, SURGICAL TAPES, AND ADHESIVES

The most common implants are the sutures. In recent years surgical tapes and tissue adhesives have added to the surgeon's armamentarium. Although their use in actual surgery is limited for some surgical procedures, they are indispensable.

9.1.1. Sutures

The types of sutures are classified by their physical integrity, i.e., absorbable and nonabsorbable. They may be distinguished according to their source of raw materials, i.e., natural sutures (catgut, silk, and cotton) and synthetic sutures (nylon, polyethylene, polypropylene, and stainless steel). Sutures may also be classified by their physical forms, i.e., **monofilament and multifilament**.

An absorbable suture, catgut, is made of collagen and derived from sheep intestinal submucosa. It is usually treated with a chromic salt to increase its strength and retard resorption by cross-linking. Such treatment extends the life of a catgut suture from 3–7 days up to 20–40 days. Table 9-1 gives some of the original strength of catgut sutures according to their sizes.

It is interesting to note that the surgical knot decreases the suture strength of catgut by half, no matter what kind of knotting technique is used, because of stress concentration. It is suggested that the most effective knotting is the square knot with three ties to prevent loosening. Whether it is tied loosely or tightly makes no measurable difference in the rate of wound healing, according to one study.

The catguts and other absorbable sutures (polyglycolic acid, PGA) invoke tissue reactions, although the effect diminishes as they are ab-

Table 9-1. Minimum Breaking Loads for British-Made Catgut[a]

Size	Diameter (mm)		Minimum breaking load (lb)	
	Minimum	Maximum	Straight pull	Over knot
7/0	0.025	0.064	0.25	0.125
6/0	0.064	0.113	0.5	0.25
5/0	0.113	0.179	1	0.5
4/0	0.179	0.241	2	1
3/0	0.241	0.318	3	1.5
2/0	0.318	0.406	5	2.5
0	0.406	0.495	7	3.5
1	0.495	0.584	10	5
2	0.584	0.673	13	6.5
3	0.673	0.762	16	8
4	0.762	0.864	20	10
5	0.864	0.978	25	12.5
6	0.978	1.105	30	15
7	0.105	1.219	35	17.5

[a] L. A. G. Rutter, "Natural Materials," in *Modern Trend in Surgical Materials*, ed. L. Gills, p. 208, Butterworths, London, 1958, by permission of the publisher.

sorbed. This is true with other natural nonabsorbable sutures like silk and cotton, which showed higher reaction than such synthetic sutures as polyester, nylon, or polyacrylonitrile, as shown in Figure 9-1. As in the wound healing process discussed in Chapter 8, the cellular response is most active one day after suturing and subsides in about a week.

If the suture is contaminated even slightly, the incidence of infection increases manyfold. The most significant factor of infection is the chemical structure; the geometric configuration seems to have no influence on infection. Polypropylene, nylon, and PGA sutures develop least infection compared to other suture materials, such as stainless steel, plain and chromic gut, and polyester sutures.

9.1.2. Surgical Tapes

Surgical tapes are supposed to offer a means of avoiding pressure necrosis, scar tissue formation, problems of stitch abscesses, and weakened tissues. The problems with surgical tapes are similar to those experienced with Band-Aids, i.e., (1) misalignment of wound edges, (2) poor adhesion

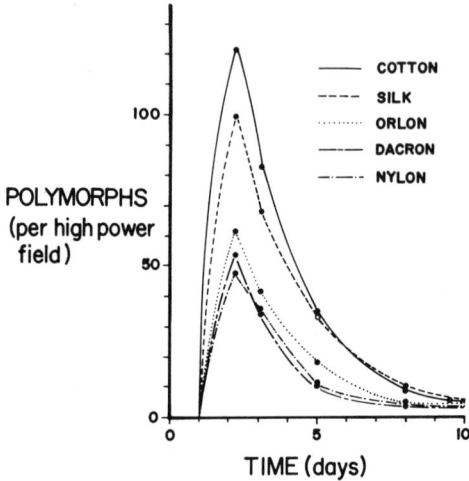

Figure 9-1. Cellular response to suture materials. (Redrawn from R. W. Postlethwait, J. F. Schaube, M. L. Dillon, and J. Morgan, *Surg., Gynecol. Obstet.*, *108*, 555, 1959, courtesy of The Franklin H. Martin Memorial Foundation.)

due to moist or dirty wound, and (3) separation of tapes when hematoma, wound drainage, etc., occur.

The wound strength and scar formation in the skin may depend on the types of incision made. If the subcutaneous muscles in the fatty tissue are cut and the overlying skin closed with tape, then the muscle retracts, which in turn increases the scar area, resulting in a poor cosmetic appearance compared to the suture closure. However, because of the greater scar tissue strength, the taped wound strength is higher than the sutured wounds if only the muscles are cut. Because of this, tapes have not enjoyed the success anticipated when they were first marketed.

Tapes have been used successfully for assembling scraps of donor skin for skin graft, connecting nerve tissues for neural regrowth, etc.

9.1.3. Tissue Adhesives

The special environment of tissues and their regenerative capacity make the development of an ideal adhesive difficult. The ideal tissue adhesive should be able to wet and bond to the tissue, be capable of rapid polymerization without excessive heat or toxic products, and be resorbable as the wound heals without interfering with normal healing processes.

The main strength of tissue adhesion comes from the covalent bond-

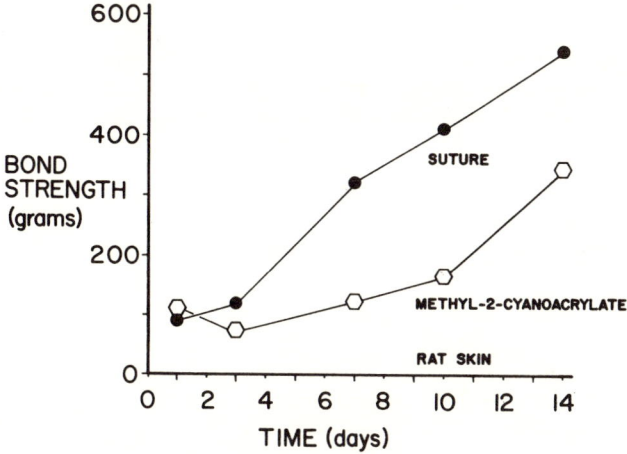

Figure 9-2. Bond strength of wounds with different closure materials. (Adapted from S. Houston, J. W. Hodge, Jr., D. K. Ousterhout, and F. Leonard, *J. Biomed. Mater. Res.*, **3**, 281, 1969.)

ing between amine, carboxylic acid, and hydroxyl groups of tissues and such functional groups as

$$R-\overset{|}{\underset{\diagdown O \diagup}{C}}-\overset{|}{C}-, \; -\overset{|}{\underset{\diagdown N \diagup}{C}}-\overset{|}{C}, \; RCNO$$
$$\underset{H}{|}$$

There are several adhesives available, of which alkyl-α-cyanoacrylate is best known. Among the homologs of alkyl-α-cyanoacrylate, the methyl- and ethyl-2-cyanoacrylate are most promising. With some plasticizers and fillers in them they are commercially known as Eastman 910 and Alpha S-2, respectively. Figure 9-2 shows that the bond strength of adhesive-treated wounds is about half that of the sutured wound after 10 days. Because of the lower strength and lesser predictability of *in vivo* performance of adhesives, the application is limited to fragile tissues after trauma, such as spleen, liver, and kidney. The topical use of the adhesives in plastic surgery and fractured teeth has been moderately successful. As with any other adhesives, the end results of the bond depend on many variables, such as thickness, porosity, and flexibility of the adhesive film, and rate of degradation.

Example 9-1

When a skin is cut in the cranial–caudal direction in the left side back of a dog after clean shaving, it opened a 2-mm gap between 1-cm cut edges. When the contralateral site was cut at 90° to the first cut, it made a 4-mm gap. Assuming the right and left sides are equivalent in properties, determine the following:

a. In which direction would you prefer to cut for surgery?
b. In which direction is the internal stress greater?
c. Once sutured, which incision will heal faster?

Answers

a. 2-mm gap direction, i.e., cranial–caudal.
b. Perpendicular to the cranial–caudal direction.
c. 2-mm gap direction.

9.2. PERCUTANEOUS AND SKIN IMPLANTS

The need for percutaneous implants has been accelerated with the advent of artificial kidneys and hearts and the prolonged injection of drugs and nutrients. Artificial skin is also badly needed to protect the body temporarily after severe burns.

9.2.1. Percutaneous Devices

The problem of obtaining a functional and a viable interface between the tissue (skin) and an implant (percutaneous device) results primarily from several factors.

1. Although an initial attachment of the tissue into the interstices of implant surface occurs, it cannot be maintained for a long time because the tissue cells continuously turn over, and downgrowth of epithelium around the implant occurs.
2. Any openings large enough for bacteria to penetrate will result in infection, even though there was an initial complete sealing of the interface between skin and implant.

Many variables and factors are involved in the development of percutaneous devices. These are:

1. End-use factors
   a. Transmission of information (biopotentials, temperature,

pressure, blood flow rate), energy (electrical stimulation, power for assist devices), and matter (cannula for blood)
   b. Transmission of load (attachment of a prosthesis).
2. Engineering factors
   a. Materials selection: polymers, ceramics, metals, and composites
   b. Design variation: button, tube with and without skirt, porous surface, etc.
3. Biological factors
   a. Implant host: man, dog, hog, rabbit, sheep, etc.
   b. Implant location: abdominal, dorsal, forearm, etc.
4. Human factors
   a. Postsurgical care
   b. Implantation technique

Selected studies performed during a fifteen-year period illustrate the difficulty of achieving the delicate balance of sealing the interface while checking the tissue growth after initial continuous attachment. They are outlined in Table 9-2.

As the table indicates, interstices for tissue ingrowth have been provided in almost all percutaneous devices. They provide a more viable connection. With few exceptions, silicone adhesives have been utilized for attaching velours or felts onto the surface of the devices. After the initial interface is formed, however, the epithelial layer cells tend to grow "downward" until they contact each other. This is nature's way of protecting itself by exteriorizing the hostile outside world. The exact mechanism of this contact inhibition, in which cells proliferate until they establish a contact, is not known.

Table 9-2. Summary of Past Studies on Percutaneous Devices

End use or purpose and design	Implant: host, location, duration	Remarks
Transmission of information (endogenous heat).	Dog, guinea pig, sheep Primate: dorsal posterior midline between scapulae; 2 weeks–9 months. *leads, Dacron felt, Silastic cast*	Good surgical technique using healthy, conditioned, parasite-free, and well-nourished animals was stressed. Claimed as a well-tolerated, simple device for chronic implantation. (K. A. Vasko and R. O. Rawson, *Trans. Am. Soc. Artif. Int. Organs*, 13, 143, 1967.)

(*continued*)

Table 9-2. (*continued*)

End use or purpose and design	Implant: host, location, duration	Remarks
Transmission of energy and control signals for an artificial heart or heart-assist device	Calves; abdomen; 12–162 days	Two-stage operation; first the prosthesis was placed under the skin with the lead extending to the body area to be serviced, second stage the skin was punctured and the tube was withdrawn to provide the passage through skin. New design was proposed to check infection. (A. Rogers and L. B. Morris, *Trans. Am. Soc. Artif. Int. Organs*, *13*, 146, 1967.)
Transmission of energy	Dog, midline of the back between the superior angles of the two scapulae; average 33 days	The lining of the percutaneous portion of the device with velour is essential to seal and prevent *downgrowth* of epithelial lining, resulting in a potentially infectible dead space. After 6 weeks the ceramic button infected and no tissue adhered to the surface. (J. Miller and C. E. Brooks, *Biomed. Mater. Res. Symp.*, No. 2, 251, 1975.)
Transmission of energy and information	Human, upper arm; 2 and 5 months	Carbon and titanium; slight inflammatory reaction, moderate epidermal downgrowth, stainless steel and gold; moderate to pronounced inflammatory reaction. Electrical impedance was fairly constant for all implants. (R. Kadefors, J. B. Reswick, and R. L. Martin, $\overline{M}ed. Biol. Eng.$, 8, 129, 1979.)
Transmission of energy and information in conjunction with attachment of prosthesis through skin	Human, leg, back; 3–6 months, 34 months longest	Four-millimeter downgrowth of the skin level was a successful implant. Failure was preceded by retardation of skin from around the neck of implant. Absence of inflammatory reaction, downgrowth of epithelium but good apposition of skin against the neck portion. Magnetic connector was used later and some

SOFT TISSUE REPLACEMENT I

Table 9-2. (continued)

End use or purpose and design	Implant: host, location, duration	Remarks
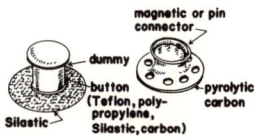		improvement was noticed since the load was not transferred through the implant. [V. Mooney and A. M. Roth, *Biomater. Med. Devices Artif. Organs*, 4(2), 171, 1976.]
Percutaneous seal; no definite purpose was placed 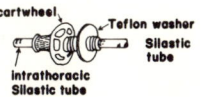	Dog, dorsal; 1, 3, 6 months	Porous high-density polyethylene is a good candidate for the percutaneous device. No difference in the acceptance by the host whether the neck portion is porous or solid. Porous neck tends to act as wick to trap bacteria. (W. E. Glazener, *Evaluation of Porous Polyethylene as a Percutaneous Seal*, M.S. thesis, Clemson University, 1975.)
Percutaneous seal, intended to be used as a percutaneous lead system	Dog, upper third of the fourth intercostal space, up to 365 days	Original idea of anchoring conduit to the ribs to provide mechanical stability was to be compromised due to the extent of periosteal bone formation. Extrusion of the Teflon conduit was caused by its rigidity, friction with the tissues, and infection. Fibroblastic activity was evident in most of tissues in contact with Teflon or Silastic. (S. Al-Nakeeb, P. T. Pearson, and N. R. Cholvin, *Biomedical Material Research*, 6, 245, 1972.)
Nonthrombogenic cannula 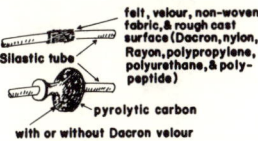	Dog, goat, pig; dorsum up to 24 months	Basal cell migration to the surface bringing the velour with it. No tissue reaction with carbon button but marsupialized, becoming explants. Growth in human is slower; hence the velour cannula is better than smooth surface, especially slightly reactive Dacron or nylon velour. (C. W. Hall, L. M. Adams, and J. J. Ghidoni, *Trans. Am. Soc. Artif. Int. Organs*, 21, 281, 1975.)

(continued)

Table 9-2. (continued)

End use or purpose and design	Implant: host, location, duration	Remarks
Cannula	Sheep: neck, back; 28 days	No significant difference between short and long cuff as an infection barrier. (J. Knight, S. J. Boyd, W. H. Van Paasschen, J. J. Cole, and B. H. Scribner, *J. Surg. Res.*, 15, 30, 1973.)
Cannula (blood access device)	Dog: neck; 20–360 days	Infection was observed once the device is pulled through skin, especially on the folding spot of the neck when the head is lowered. (T. McEvoy, R. Wathen, R. Forstron, F. Dorman, and B. Monson, *Trans. Am. Soc. Artif. Int. Organs*, 21, 289, 1975.)
Cannula	Dog: abdominal, subcutaneous Rabbit: transcutaneous; 2, 4, 6, 27, and 40 weeks Man: peritoneal dialysis patients; 2–18 weeks	Ivalon sponge calcified after 27 and 70 weeks of subcutaneous implant. The transcutaneous implant with Dacron cuff showed slight inflammatory reaction but after 5 weeks downgrowth of epithelium was evident. (G. E. Striker and H. A. M. Tenckhoff, *Surgery*, 69, 70, 1971.)
Cannula (grafted blood access device)	Dog: renal artery subcutaneous, extracutaneous, detachable Sheep: neck extracutaneous; 16.8 days average	No observation was made about the tissue and cannula interface because the investigators are only interested in the development of nonthrombogenic device. (Y. Nósé et al., Proc. 7th Ann. Contrac. Conf., Art. Kidney–Chronic Uremia Prog., NIAMDD, Washington, D.C., 1974.)
Cannula	Man: arm; 5 weeks–17 months	No observation was made about the tissue and cannula interface because the investigators are only interested in the development of nonthrombogenic device.

SOFT TISSUE REPLACEMENT I

Table 9-2. (*continued*)

End use or purpose and design	Implant: host, location, duration	Remarks
	Dacron velour / Hepacone	(C. A. Hufnagel, M. N. Gomes, and D. Ozdemir, Proc. 6th Ann. Contrac. Conf., Art. Kidney–Chronic Uremia Prog., NIAMDD, Washington, D.C., 1973.)
Cannula	Rabbit: dorsal flank; 6–119 days	Ingrowth of tissue into clipped velour produced a strong bond which will not permit the epidermis to slide outward as it is exfoliated. Thus, the epidermal portion of the button should be smooth. Ingrowth into looped velour and open-mesh Dacron skirt materials produces a strong bond, which fails by the plastic filaments cutting through the tissue; thus thicker and soft-coated separators of the ingrowth openings are desirable. The modulus of even the "soft" Silastic button is still too high to match that of skin. (F. W. Cooke et al., Final Report, NIAMDD, Contract No. NIH-69-83, Washington, D.C., 1971.)

9.2.2. Artificial Skin

Artificial skin is another example of percutaneous implants and has similar problems. Most needed for this application is a material that can adhere to a large (burned) skin surface and prevent the loss of fluid, electrolytes, and other biomolecules until the wound is healed. Although a permanent skin implant is needed, it is a long way from being developed because of the same reasons given for percutaneous devices proper. Autografting and homografting are the only methods presently available.

Several polymeric materials, including reconstituted collagen, have been tried. Among them are the copolymers of vinyl chloride and acetate and methyl-2-cyanoacrylate. The methyl-2-cyanoacrylate was found to be too brittle and histotoxic for use as a burn dressing. The ingrowth of tissue into the pores of sponge (Ivalon®, polyvinyl alcohol) and woven fabric (nylon and silicone rubber velour) was attempted without much success. Sometimes plastic tapes were used to hold skin grafts during microtombing and grafting procedures. For severe burns, immersion into silicone fluid was found to be beneficial for preventing early fluid loss, decubitus ulcers, and reducing pain.

Example 9-2

A bioengineer is trying to understand the biomechanics of a hole created in the skin for a transcutaneous implant. He made a hole using a circular biopsy drill in the dorsal skin of a dog. The diameter of the drill is 5 mm. If the hole became an ellipse with a minor and major axis of 3 and 7 mm, answer the following questions.

a. In which direction is the internal stress of the skin greater?
b. In which direction are the collagen fibers more oriented?
c. How can the bioengineer obtain a circular hole rather than an ellipse for the implant?
d. Assuming the implant is nondeformable compared to the skin, what problems will arise between skin and implant when a load or force is applied to the skin or implant by handling or accident?

Answers

a. In the major axis.
b. In the major axis.
c. By making an elliptic hole with a major and minor axis in the opposite of the ellipse made by the circular hole.
d. All the load will be transferred through the implant and the deformation will be stopped at the interface, which will result in a large deformation at the interface. It is likely that the interface will fail if it cannot absorb the deformation.

9.3. MAXILLOFACIAL AND OTHER SOFT TISSUE AUGMENTATION

The previous section dealt with problems associated with wound closing and wound-tissue interfacial implants. This section will consider (cosmetic) reconstructive implants. Although soft tissue implants can be divided into (1) space fillers, (2) mechanical supports, and (3) fluid carriers or storers, most of them have two or more combined functions. For example, breast implants fill space and provide mechanical support.

9.3.1. Maxillofacial Implant

There are two types of maxillofacial implant materials (often called prosthetics, which implies extracorporeal attachment): extraoral and intraoral. The latter is implanted and the former is not. Maxillofacial implant is defined as "the art and science of anatomic, functional or cosmetic reconstruction by means of artificial substitutes of those regions in the maxilla, mandible, and face that are missing or defective because of surgical intervention, trauma, etc." (V. A. Chalian and R. W. Phillips, *J. Biomed. Mater. Res. Symp.*, 5, 349, 1973).

Many polymeric materials are available for the extraoral implant, which requires that (1) color and texture should be matched with that of patients; (2) it should be mechanically and chemically stable, i.e., it should not creep and change colors or irritate skin; and (3) it should be easily fabricated. Polyvinyl chloride and acetate (5–20%) copolymers, polymethylmethacrylate, silicone, and polyurethane rubbers are currently used.

The requirements for the intraoral implants are the same as for other implant materials. For the maxilla, mandibular, and facial bone defects, metallic materials such as tantalum and Vitallium® are used. For soft tissues like gum and chin, such polymers as silicone rubber and polymethylmetyacrylate are used for augmentation.

The use of injectable silicones that polymerize *in situ* was partially successful for correcting facial and other deformities. Although this is obviously a better approach in view of the minimal surgical damage initially, this procedure was not accepted because of tissue reaction and eventual implant displacement. It has now been abandoned.

9.3.2. Other Soft Tissue Implants

Breast implants are quite common. In early stages the enlargement of breasts was done with various materials such as paraffin wax, beeswax, or silicone fluids by direct injection or enclosed in a rubber balloon. Several problems have been associated with the directly injected implants, including progressive instability, ultimate loss of original shape and texture, infection, and pain. In the late 1960s the FDA banned such practices by classifying as drugs such injectable implants as silicone gel.

Another effort at augmenting breast size was to implant a sponge made of polyvinyl alcohol. However, the ingrown soft tissues calcified with time and so-called marble breast resulted. Although the enlargement or replacement of breast for cosmetic reasons alone is not recommended, prostheses have been developed for patients who have undergone radical mastectomy or who have nonsymmetrical deformities. They are probably beneficial for psychological reasons. In this case the silicone rubber bag filled with silicone gel and backed with polyester mesh to permit tissue ingrowth for fixation is widely accepted. The artificial penis, testicles, and vagina fall into the same category as breast implants.

PROBLEMS

9-1. A cut skin is to be closed by using either a suture or a surgical adhesive, methyl-2-cyanoacrylate. Which of the two can be removed earlier?

Answer
A sutured wound will heal faster and the suture can be removed earlier than surgical adhesive could be (see Figure 9-2).

9-2. The bioengineer excised out the skin of example 9-2 with and without the hole and put on a tensile testing machine (the width and gage length of the samples were 4 cm and 4 mm, respectively).

a. If the average forces recorded after straining 30% for the samples with and without the hole were 100 and 160 N, what are the stresses?
b. Calculate the percent stress decrease by making the hole.
c. Indicate where the stress is the greatest in the sample with the hole.
d. If the hole is replaced with a rigid implant with a good interface, would the stress be greater than an empty hole? Assume that other factors are the same.

Answers

a. $$\sigma = \frac{f}{A} = \frac{160 \text{ N}}{40 \times 4 \times 10^{-6} \text{ m}^2} = \underline{1 \text{ MPa}}$$

$$\sigma_{hole} = \frac{100 \text{ N}}{35 \times 4 \times 10^{-6} \text{ m}^2} = \underline{0.71 \text{ MPa}}$$

b. $$\frac{1 - 0.71}{1} = \underline{0.29 \ (29\%)}$$

c.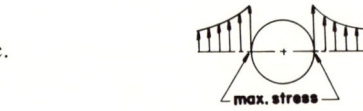

d. Greater

SOFT TISSUE REPLACEMENT I

9-3. List the order of preference for use in a transcutaneous implant.

a. Silicone rubber
b. Polymethylmethacrylate
c. Polyethylene
d. Polyurethane
e. Stainless steel
f. Carbon

Answer
a, d, c, b, f, and e.

9-4. It is suggested that the stability of hydrolytes of a polymer can give a good estimate of potential stability *in vivo*. Explain why and give examples.

Answer
Hydrolyzation of the chain molecules of a polymer will severe the chains, resulting in chain scission. Polymers with —HHCO— and —COO— linkages in their main chain, such as polyurethane, polyurea, polyester (except polyethylene terephtalate, Dacron®), and polyamide (see Table 6-1) showed deterioration *in vivo*.

9-5. Suppose that you are a bioengineer trying to develop a new artificial skin. Suggest steps to ensure that the new product will be used widely in clinics.

9-6. List the problems associated with percutaneous devices and suggest ways of improving them.

Answer
As discussed in the text, the major problem is infection and the second is epithelial downgrowth. Because the implant-skin interface is the key to the problem, improvements should come from better understanding of the interface. First, the problem of trauma can be attacked by understanding the stress applied during accidental handling of the implant. The hole made to place the implant acts as a stress riser, which warrants a systematic study. It is also speculated that electrical stimulation and retardation can be applied in the implant, especially at the interface. Initially, the electric current will be applied for rapid growth of the dermis, reducing the healing period; later, by changing the polarity, retarded tissue growth will prevent epithelial downgrowth. Other innovations, such as grafting mucosa around the implant for immediate fixation of the implant and delayed later-stage skin growth, are being attempted.

FURTHER READING

A. H. Bulbulian, *Facial Prosthetics*, Charles C Thomas, Springfield, Ill., 1973.
V. A. Chalian, J. B. Drane, and S. M. Standish, *Maxillofacial Prosthetics,* Williams & Wilkins Co., Baltimore, 1971.
S. F. Hulbert, S. N. Levine, and D. D. Moyle (eds.), *Prosthesis and Tissue: The Interface Problem,* pp. 99–136, J. Wiley and Sons, New York, 1974.
H. Lee and K. Neville, *Handbook of Biomedical Plastics,* chapters 4 and 13, Pasadena Technology Press, Pasadena, Calif., 1971.

S. N. Levine (ed.), "Polymers and Tissue Adhesives," in *Ann. N.Y. Acad. Sci.*, Part IV, *146*, p. 193, 1968.

V. Mooney, S. A. Schwartz, A. M. Roth, and M. J. Gorniowsky, "Percutaneous Implant Devices," *Ann. Biomed. Eng.*, *5*, 34, 1977.

G. D. Winger, "Epidermal Regeneration Studied in the Domestic Pig," in *Epidermal Wound Healing,* chapter 4, ed. H. I. Maibach and D. T. Rovee, Year Book Medical Publishers, Chicago, 1972.

CHAPTER 10

SOFT TISSUE REPLACEMENT II: BLOOD INTERFACING IMPLANTS

The blood interfacing implants can be divided into two categories: (1) short-term extracorporeal implants such as membranes for artificial organs (kidney and heart–lung machine) and tubes and catheters for the transport of blood; and (2) long-term *in situ* implants such as vascular implants and implantable artificial organs. Although pacemakers for the heart are not interfaced with blood they are considered here because these devices help to circulate blood throughout the body.

The single most important requirement for the blood interfacing implants is blood compatibility. Although blood coagulation is the most important factor for compatibility, the implants should not damage proteins, enzymes, and formed blood elements (red blood cells, white blood cells, and platelets). The latter includes hemolysis (red blood cell rupture) and initiation of the platelet-release reaction.

Coagulated blood is called a clot. However, sometimes the clot formed inside the blood vessels is referred to as thrombus or embolus, depending on whether the clot is fixed or floating.

The mechanism and route of blood coagulation are not completely understood. A simplified version of blood clotting is proposed as a cascading sequence, as shown in Figure 10-1. As discussed in Section 8.1, immediately after injury the blood vessels constrict to minimize the flow of

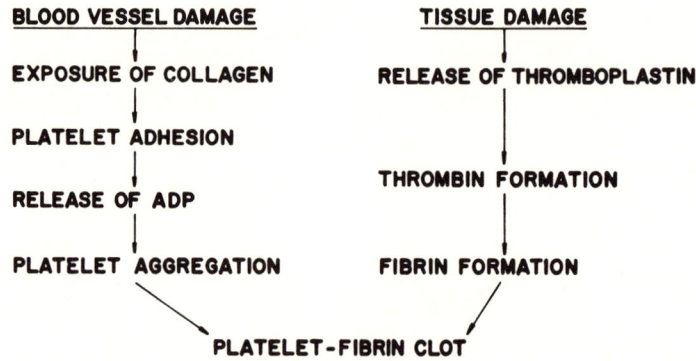

Figure 10-1. Two routes for blood clot formation (note the cascading sequence).

blood. Platelets adhere to the vessel walls by contacting with the exposed collagen. The aggregation of platelets is achieved through release of adenosine diphosphate (ADP) from damaged red blood cells, vessel walls, and adherent platelets.

Clot formation is achieved by fibrin deposits around the platelet aggregates. The formation of fibrin is due to the thrombin, which in turn is produced by the thromboplastin released through damaged tissues including the vessel walls. Sometimes the blood coagulation in the absence of thromboplastin is called the "intrinsic" route.

Any interruption of the cascade sequence in Figure 10-1 can prevent the fibrin formation that precedes blood coagulation. Anticoagulants such as heparin complexes with plasma thromboplastin thus prevent any fibrin formation. The plasma thromboplastin cannot be activated without Ca^{2+} ions; hence removing calcium by chelating with citrate and oxalate gives the same result.

10.1. BLOOD COMPATIBILITY

Materials used for interfacing with blood can be divided into short- and long-term groups: the former includes catheters and inner linings of heart-lung machines, oxygenators, and kidney dialyzers; the latter includes vascular implants, heart valves, etc. Many factors affect the blood compatibility, mainly related to the formation of blood clots (or thrombi). Clot formation is largely involved with the nature of the surface interfacing with the blood.

10.1.1. Factors Affecting Blood Compatibility

Surface roughness is an important factor because the rougher the surface the more area is exposed to blood. Therefore, the rough surface promotes faster blood coagulation than the highly polished surface of glass, polymethylmethacrylate, polyethylene, and stainless steel. Sometimes thrombogenic (clot-producing) materials with rough surfaces are used to promote clotting in porous interstices to prevent initial leaking of blood and later tissue ingrowth through the pores of vascular implants.

The surface wettability, i.e., its hydrophilic (wettable) or hydrophobic (nonwettable) nature, was once thought an important factor. However, the wettability parameter, i.e., contact angle with liquids, does not correlate consistently with clotting characteristics.

The surface of intima of blood vessel is negatively charged ($1 \sim 5$ m V) with respect to the adventitia. This phenomenon is attributed partially to the nonthrombogenic or thromboresistant character of the intima, since the formed elements of blood are also negatively charged and hence are repelled from the surface of the intima. This was demonstrated by using a solid copper tube, which is a thrombogenic material, implanted as an arterial replacement. When the tube was negatively charged, the clot formation was delayed when compared with the control. In relation to this phenomenon, the streaming or zeta potential has been investigated because the formed elements of blood are flowing particles *in vivo*. However, it was not possible to establish a direct one-to-one relationship between clotting time and zeta potential.

The chemical nature of the material surface interfacing with blood is closely related to the electrical nature of the surface because the type of functional groups of polymer determines the type and magnitude of the surface charge. (No intrinsic surface charge exist for metals and ceramics, although some ceramics and polymers can be made piezoelectric.) The surface of intima is negatively charged largely because of the presence of mucopolysaccharides, especially chondroitin sulfate and heparin sulfate.

10.1.2. Nonthrombogenic Surfaces

Many studies have sought nonthrombogenic materials, often using the empirical approach. These materials can be categorized as (1) heparinized or biologic surfaces, (2) surfaces with anionic radicals for negative electric charges, (3) inert surfaces, and (4) solution-perfused surfaces.

Heparin is a polysaccharide with negative charges due to the sulfate groups:

$$\left[\begin{array}{c} \text{CO}_2\text{H} \\ \text{OH} \\ \text{OH} \end{array} \underset{O}{\diagdown} \begin{array}{c} \text{CH}_2\text{OH} \\ \text{OH} \\ \text{NH}_2 \end{array} \underset{O}{\diagdown} \begin{array}{c} \text{CO}_2\text{H} \\ \text{OH} \\ \text{OH} \end{array} \underset{O}{\diagdown} \begin{array}{c} \text{CH}_2\text{OH} \\ \text{OH} \\ \text{NH}_2 \end{array} \right]_n + 5 \text{ sulfate groups} \quad (10\text{-}1)$$

Initially the heparin was attached to the graphite surface treated with quaternary salt, benzalkonium chloride (GBH process). Later a simpler heparinization was accomplished by exposing the polymer surface to a quaternary salt, such as tridodecylmethylammonium chloride (TDMAC). This method was further simplified by making TDMAC and heparin solution in which the implant can be immersed followed by drying.

The heparinized materials showed a significant increase in thromboresistance compared to an untreated control. In an interesting application, a polyester fabric graft was heparinized. This reduced the tendency of initial bleeding through the fabric and a thin neointima was later formed. Many polymers were tried for heparinization, including polyethylene and silicone rubber. Leaching of the heparin into the medium is a drawback, although some improvement was seen by cross-linking of the heparin with gluteraldehyde and directly covalent-bonding it onto the surface.

In some studies, the cardiovascular implant surface was coated with other biological molecules such as albumin, gelatin (denatured collagen), and heparin. Some reported that the albumin alone can be thromboresistant and decrease the platelet adhesion.

Surfaces with anionic radicals or negative electric charges were made by copolymerization or grafting the surface with anionic radicals. Some experimental results of this approach are given in Table 10-1. It has also been shown that negatively charged electretes on the surface of a polymer enhance thromboresistance.

Hydrogels of both hydroxyethylmethacrylate (poly-HEMA) and acrylamide are classified as inert materials because they neither contain highly negative anionic radical groups nor are negatively charged. Like the heparin coatings, these coatings tend to be washed away when exposed to the bloodstream. Segmental or block polyurethanes are also somewhat thromboresistant without surface modification.

Another method of making surfaces nonthrombogenic is by perfusion of water (solution) through interstices of fabric that interface with blood. This new approach has the advantage of avoiding damages of formed elements and is nonthrombogenic. The disadvantage is the dilution of blood plasma although this problem is not serious because saline solution is deliberately injected for the kidney and heart–lung machine.

Table 10-1. Relationship between Thrombogenicity of Surfaces and Fibrinogen on Surfaces[a]

Material	Fibrinogen on surface (μg/cm^2)		Nonthrombogenic character[d]
	Picked up[b]	Adsorbed[c]	
Glass	3.14, 2.37	0.119	Very poor, minutes[e]
Polytetrafluoroethylene (PTFE)	0.28	0.156	Fair, days to weeks
Etched PTFE	1.02, 2.58	0.465	Poor to good
Hydroxyl-coated PTFE	2.18	0.471	Poor, days to weeks
Carboxyl-coated PTFE (Van de Fraaff)	2.11, 2.58	0.308, 0.362	Poor
PTFE coated with sulfated hydroxyl	2.16, 2.02	0.460, 0.326	Fair, weeks
Ethylenesulfonic-coated	1.61	0.542	Fair, weeks
Lightly sulfonated PTFE	1.15	0.472	Fair to good, weeks to months
Highly sulfonated PTFE	1.31, 2.89	0.625, 0.647	Very good, months

[a] From T. F. Ziegler and M. L. Miller, *J. Biomed. Mater. Res.*, 4, 259, 1970; courtesy of J. Wiley and Sons.
[b] This figure includes undenatured fibrinogen adsorbed from the underlying solution, which contained 0.05 to 0.10 mg of fibrinogen per 100 ml.
[c] From solution containing 0.33 mg of fibrinogen in 100 ml of water.
[d] Cannulae, having the surfaces under test, we implanted in the blood vessels of dogs and the length of time the dogs lived was used as a measure of the thrombogenicity.
[e] Based on *in vitro* tests.

Example 10-1

In circulation of the blood, the formed elements are being destroyed by the blood pump and tube wall contacts. A bioengineer measured the rate of hemolysis (red blood cell lysis) as 0.1 g/100 liters pumped. If the normal cardiac output for a dog is 0.1 liter/kg/min, what is the hemolysis rate? If the animal weighs 20 kg and the critical amount of hemolysis is 0.1 g/kg of body weight, how long can the bioengineer circulate the blood before reaching a critical condition? Assume a negligible amount of new blood formation.

Answer

Hemolysis rate = 0.1 g/100 liters (0.1 liter/kg/min) × 20 kg = <u>2 mg/min</u>

Critical hemolysis = 0.1 g/kg × 20 kg = <u>2 g</u>

$$\frac{2 \text{ g}}{2 \text{ mg/min}} = \underline{1000 \text{ min}} \text{ (or 16 h 40 min)}$$

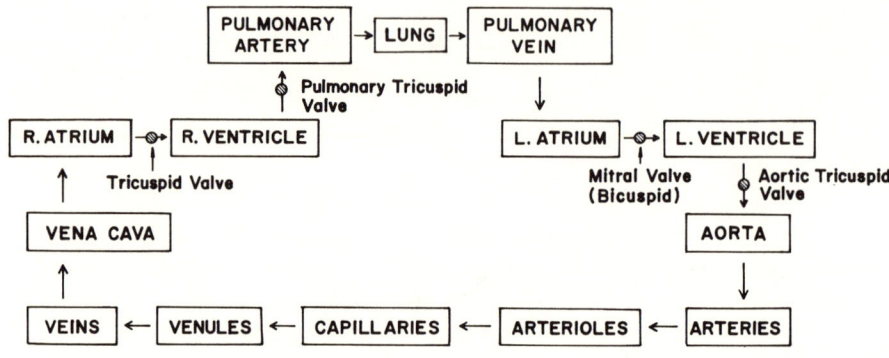

Figure 10-2. Schematic diagram of blood circulation in the body.

10.2. IMPLANTS FOR BLOOD INTERFACE

Blood is circulated throughout the body according to the sequence shown in Figure 10-2. Implants are usually used to replace or patch large arteries and veins, including the heart and its valves, although surgical remedy without using implants is usually preferred. However, there are many unavoidable occasions when it is necessary to anastomose a large segment of the vital organs with implants.

The basic requirements for the blood-interfacing implants are the same as other soft tissue implants (Chapter 9), except the surface exposed to blood should be made nonthrombogenic or at least thromboresistant. Most materials used for this type of applications are made of polymers because of their flexibility and ease of fabrication.

10.2.1. Vascular Implants

Implants have been used in various vascular maladies, ranging from simple sutures for anastomosis after removal of vessel segments to patches for aneurysms. The vein implant encountered some difficulties due to the collapse of the adjacent vein or clot formation because of low pressure and stagnant flow. Vein replacements have not been a major concern because autografts can be performed in most cases. Nonetheless, many materials, including nylon, polytetrafluoroethylene, and polyester, have been fabricated for clinical applications.

Early designs for arterial implants were solid tubes made of glass, aluminum, gold, silver, and polymethylmethacrylate. All implants developed clots. In the early 1950s porous implants were introduced that allowed tissue growth into the interstices. The new tissues interface blood and thus minimize clotting. Ironically, thrombogenic materials were found satisfactory for this type of application. Another advantage of tissue ingrowth is the fixation of implant by the ingrown tissue, making a viable anchor. The initial leakage through pores is disadvantageous but this problem can be prevented by preclotting the outside surface of the implant prior to placement.

Although the exact sequence of the formation of tissue in the human implants is not fully documented, quite a bit is known about reactions in animals. Generally, soon after implantation the inner and outer surfaces of the implant are covered with fibrin and fibrous tissue, respectively. A layer of fibroblasts replaces the fibrin, becoming neointima (sometimes called pseudointima). The long-term fate of the neointima varies with animals; in the dog it stabilizes into a constant thickness; for the pig it will grow until it occludes the vessel. In humans the initial phase of the healing is the same as in animals but in latter stages the inner surface is covered by both fibrin and a cellular layer of fibroblasts. The sequence of healing of arterial implants is given in Figure 10-3.

Types of material and the geometry of the implant influence the rate and nature of tissue ingrowth. Several polymer materials are currently used to fabricate implants, including nylon, polyester, polytetrafluoroethylene, polypropylene, and polyacrylonitrile. However, polytetrafluoroethylene,

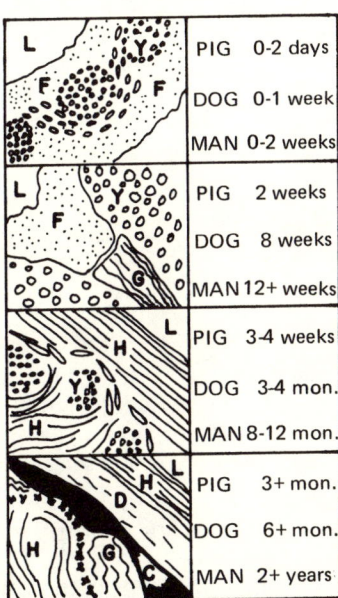

Figure 10-3. Basic healing pattern of arterial prothesis. L = lumen of prothesis, F = fibrin, Y = yarn bundle, G = organizing granulation tissue, H = healed fibrous capsular tissue, D = degenerative fibrous capsular tissue, C = calcified capsular tissue. (Redrawn from S. A. Wesolowski, C. C. Fries, A. Martinez, and J. D. McMahon, *Ann. N.Y. Acad. Sci.,* 146, 325, 1968.)

polyester, and polypropylene are the most favorable materials due to their minimal deterioration of physical properties *in vivo* as discussed in Chapter 6. Polyester (particularly polyethyleneterephthalate, Dacron®) is usually preferred because of its superior handling properties.

The geometry of fabrics and resulting porosity have a great influence on healing characteristics. The preferred porosity is such that 5000 and 10,000 ml of water are passed per 1 cm² of fabric per minute at 120 mm Hg. The lower limit is to prevent excessive leakage of blood and the higher limit is for better tissue ingrowth and healing. Implant thickness is directly related to the amount of thrombus formation; the thinner the fabric the smaller or the thinner the thrombus deposit and the faster the organization of the neointima.

Example 10-2

a. Calculate the maximum tension developed for an artery with a 0.5-cm diameter. Assume that the maximum pressure will be 250 mm Hg and that the artery is uniform in length.
b. To replace a section of artery of about 5 cm, what is the maximum force exerted on the wall?
c. Can silicone rubber be used for the replacement material if the wall thickness is 1 mm and the safety factor is 10?

Answers

a. From equation (7-8)

$$T = P \cdot r$$
$$= 250 \text{ mm Hg} \cdot 1.33 \times 10^2 \text{ Pa/mm Hg} \cdot 0.25 \text{ cm}$$
$$= \underline{83 \text{ N/m}}$$

b. $F = 83 \dfrac{N}{m} \times 5 \times 10^{-2} \text{ m} = \underline{4.2 \text{ N}}$

c. $\sigma = \dfrac{F}{A} = \dfrac{4.2 \text{ N}}{5 \times 10^{-2} \text{ m} \times 1 \times 10^{-3} \text{ m}} = \underline{0.84 \times 10^5 \text{ Pa}}$

Since the safety factor is 10, the maximum wall stress should be 0.84×10^6 Pa or 0.84 MPa. From Table 6-4 the tensile strength of silicone rubber is 6 MPa, which is adequate for this purpose; however, it is not used because of its inadequate cyclic fatigue property and absorption of small molecules *in vivo*.

10.2.2. Heart Valve Implants

The four valves in the ventricles of the heart are shown in Figure 10-4 (cf. Fig. 10-2). In the majority of cases, valves in the left ventricle (mitral

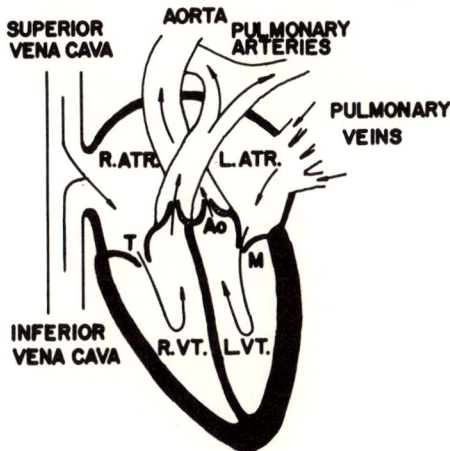

Figure 10-4. Circulation of blood in the heart. (Compare with Fig. 10-2.)

and aortic) become incompetent because of high pressure. The most important and frequently critical is the aortic valve, which is the last gate the blood has to go through before being circulated in the body.

There have been many different types of valve implants. The early ones in the 1960s were made of leaflets that mimicked the natural valves. Invariably, the leaflets could not withstand fatigue for more than three years. In addition to hemolysis, regurgitation and incompetence were problems. Later, butterfly leaflets and ball or disk-in-the-cage valves were introduced, some of which are shown in Figure 10-5. The material requirements for valve implants are the same as for vascular implants. Some additional requirements are related to blood flow and pressure, i.e., the formed elements of blood should not be damaged and should not drop the blood pressure below clinically significant values. Also, noise should be minimal.

All valves have a sewing ring that is covered with various polymeric fabrics. This helps the initial fixation of implant and later the tissue ingrowth will render the fixation viable similar to the porous vascular implant. The cage itself is usually made of metals and covered with fabrics for reducing noise or with pyrolytic carbon for a nonthrombogenic surface (the disk or ball is also coated with pyrolytic carbon at the same time).

The ball (or disk) is made of hollowed solid polymers (polypropylene, polyoxymethylene, polychlorotrifluoroethylene, etc.), metals (titanium, Co-Cr alloy), and pyrolytic carbon. The early use of silicone rubber poppet was found undesirable because of lipid adsorption and subsequent swelling and dimensional changes. Although this was an unfortunate

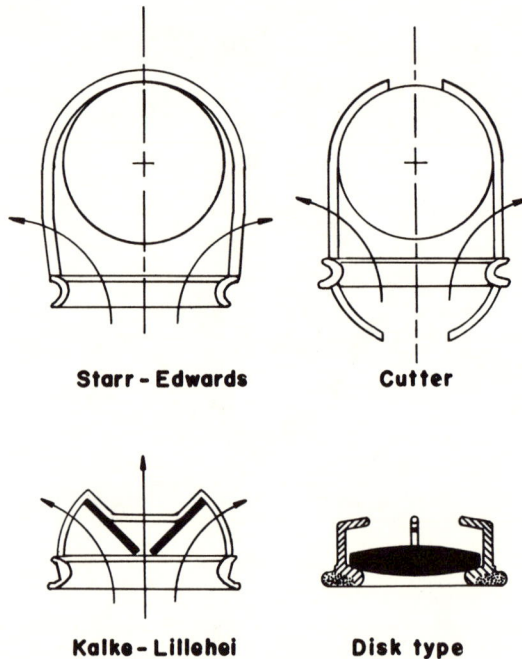

Figure 10-5. Various types of heart valves.

episode (some implants were fatal), it helped to reinforce that the *in vitro* experiment alone is not sufficient to predict all the circumstances arising from *in vivo* use no matter how carefully predictions are made. This is true of any implant, even a very simple device.

10.2.3. Heart Assist Devices

Heart assist devices are aimed at sustaining blood circulation when the natural heart cannot function normally or during cardiac surgery. Though the blood can be circulated by a pump it can be oxygenated either by the patient's own lung or through an artificial oxygenator. Often the latter method is preferred for the simplicity of the operation during surgery.

The three basic types of oxygenators are shown in Figure 10-6. In all cases oxygen gas is contacted with blood and simultaneously waste gas (CO_2) is removed. To increase the rate of gas exchanges at the blood-gas surface of the bubble oxygenator, the gas is broken into small bubbles (about 1-mm diameter; if too small, they are hard to remove from blood), increasing the surface contact area. Sometimes the blood is spread thinly as a film and exposed to the oxygen; this is called a film oxygenator. The membrane oxygenator is similar to the artificial kidney membrane discussed later. The main difference is that the oxygenator membrane is permeable to gases only while the kidney membrane is permeable to liquids.

Some of the mechanical and chemical characteristics of the natural and artificial lung (oxygenator) are given in Table 10-2. The surface area of the artificial membrane is about 10 times larger than the natural lung since the amount of O_2 transfer through a membrane is proportional to the surface area, pressure, and transit time, but inversely proportional to the film thickness (blood). Because the blood film thickness of the artificial membrane is about 30 times larger than the natural lung, it must compensate by increased transit time (16.5 s) and higher pressure (650 mm Hg) to provide the same amount of O_2 transfer as the lung.

The membranes are usually made of silicone rubber or poly-

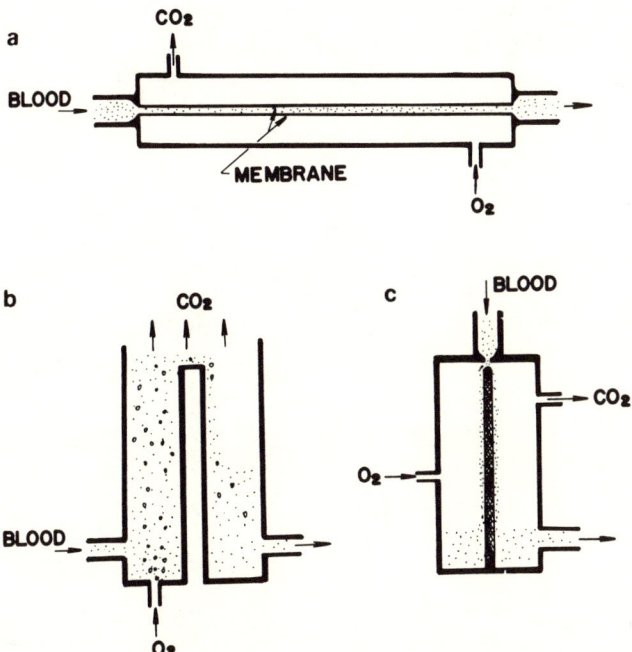

Figure 10-6. Oxygenators: (a) membrane (b) bubble, and (c) film.

Table 10-2. Mechanical and Chemical Characteristics of Natural versus Artificial Lung[a]

Function	Natural lung	Artificial lung
Pulmonary flow	5 liters/min	5 liters/min
Head of pressure	12 mm Hg	0–200 mm Hg
Pulmonary blood volume	1 liter	1–4 liters
Blood transit time	0.1–0.3 s	3–30 s
Blood film thickness	0.005–0.010 mm	0.1–0.3 mm
Length of capillary	0.1 mm	2–20 cm
Pulmonary ventilation	7 liters/min	2–10 liters/min
Exchange surface	50–100 m^2	2–10 m^2
Venoalveolar O_2 gradient	40–50 mm Hg	650 mm Hg
Venoalveolar CO_2 gradient	3–5 mm Hg	30–50 mm Hg

[a] Adapted from D. O. Cooney, *Biomedical Engineering Principles*, p. 404, Marcel Dekker, New York, 1976.

tetrafluoroethylene. The gas permeability of these materials is given in Table 10-3. Silicone rubber is 40 and 80 times more permeable to O_2 and CO_2 than the polytetrafluoroethylene, but the latter can be made about 20 times thinner. Therefore, silicone rubber is only 2 and 4 times better for O_2 and CO_2 transfer than polytetrafluoroethylene.

Polyurethane and natural and silicone rubber have been used for constructing balloon-type assist devices as well as for coating the inner

Table 10-3. Gas Permeability of Teflon and Silicone Rubber Membranes[a]

Membrane	Thickness (10^{-3} m)	Oxygen	Carbon dioxide	Nitrogen	Helium
Teflon	⅛	239	645	106	1425
	¼	117	302	56	730
	⅜	77	181	35	430
	½	61	126	30	345
	¾	41	86	23	240
	1	29			
Silicone rubber	3	391	2072	184	224
	4	306	1605	159	187
	5	206	1112	105	133
	7	159	802	81	94
	12	93	425	48	51
	20	59	279	31	43

[a] Permeation rates of oxygen, carbon dioxide, nitrogen, and helium across Teflon and silicone rubber membranes of a given thickness, in ml/min · m^2 · atm (STP).

SOFT TISSUE REPLACEMENT II

surfaces of total artificial hearts. These materials are thromboresistant. Sometimes the surfaces are coated with heparin and other nonthrombogenic molecules. Felt or velour surfaces were not successful as imitations of natural tissue on the inner surface of an artificial heart wall.

10.2.4. Artificial Organs

The ultimate triumph of implant science would be to make implants that behave or function exactly the same as the organs or tissues they replace. As mentioned in Chapter 1, most implants are designed to substitute for mechanical functions. Electrical functions can be taken over by implants (pacemakers), and some primitive yet vital chemical functions can also be delegated to implants (kidney machine and oxygenator).

Most artificial heart and heart assist devices use a simple balloon and valve system. Figure 10-7 shows some typical heart assist devices. In all cases a balloon is used to displace blood. A simpler heart assist device is the intraaortic balloon, which is placed on the descending aorta. During the diastolic phase of the heart the balloon is inflated to prevent backflow.

Two artificial hearts are shown in Figure 10-8. Although their design principles and material requirements are the same as those for assist devices, the power consumption ($\sim$6 watts) is too high to be implanted completely at this time. A miniature, totally implantable, nuclear-powered, artificial heart is being developed. So far the power is introduced through a percutaneous device (Chapter 9) in the form of compressed air or electricity.

A cardiac pacemaker is used to assist the regular contraction rhythm of heart muscles. The sinoatrial (SA) nodes of the heart originate the electri-

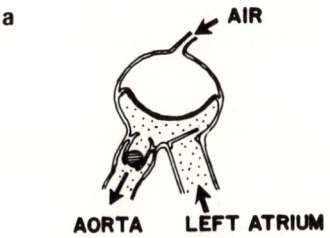

Figure 10-7. Heart assist devices. (a) De Bakey left ventricular bypass. (b) Bernhard-Teco assist pump. (Redrawn from H. Lee and K. Neville, *Handbook of Biomedical Plastics*, pp. 6–24, Pasadena Technology Press, Pasadena, Calif., 1971.)

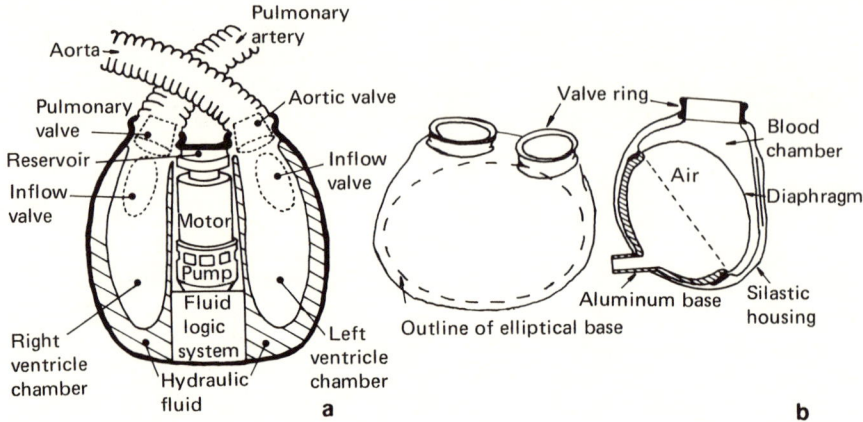

Figure 10-8. Artificial hearts. (a) Schumacker–Burns electrohydrolic heart (H. Lee and K. Neville, *Handbook of Biomedical Plastics,* pp. 6–32, Pasadena Technology Press, Pasadena, Calif., 1971). (b) Jarvik-III-type artificial heart (W. J. Kolff, *Artificial Organs,* p. 18, J. Wiley and Sons, New York, 1976).

cal impulses that pass through the bundle of His to the atrioventricular (AV) node. In the majority of cases pacemakers are used to correct the conduction problem in the bundle of His. Basically, pacemakers should deliver an exact amount of electrical stimulation to the heart at varying heart beats. The pacemakers consist of conducting wires with a power source, as shown in Figure 10-9.

The wires are well insulated with rubber (usually silicone) except the tips, which are sutured or embedded directly in the cardiac wall. The tip is usually made of a noncorrosive noble metal with reasonable mechanical strength such as Pt–10% Ir alloy. The most significant problems are the fatigue of the wires (they are coiled like a spring to prevent this) and the collagenous scar tissue formation at the tip, which increases the electrical resistance at the junction. The mercury battery and electronic circuits are insulated by casting with a clear polymer resin.

The pacemakers are usually changed between 2 and 5 years because of the limit of the power source. A nuclear energy-power pacemaker is commercially available. Although this new power pack may lengthen the life of the power source, the fatigue of the wire and diminishing conductivity caused by the thickening of tissue will limit its maximum lifetime to less than 10 years. A porous electrode at the tip of the wire may be fixed to the cardiac muscle by tissue ingrowth, as in vascular prostheses, which may solve the interfacial problem.

The primary function of the kidneys is to remove metabolic waste

SOFT TISSUE REPLACEMENT II 177

products by passing blood through a glomerulus under a pressure of about 75 mm Hg. The glomerulus contains up to 10 primary branch and 50 secondary loops to filter the blood. The glomeruli are contained in the Bowman capsule, which in turn is a part of the nephron (Fig. 10-10). The main filtrates are urea (70 times the urea content of normal blood), sodium, chloride, bicarbonate, potassium, glucose, creatinine, and uronic acid. About 120 ml is filtered per minute, of which 2 ml is excreted and the rest is reabsorbed.

The membrane is the key component of an artificial kidney machine. In fact, the first attempt to filter or dialyze blood with a machine failed because of an inadequate membrane. Besides a membrane filter the kidney dialyzer consists of a bath and a pump to circulate blood from an artery and return the cleansed blood to the vein, as shown in Figure 10-11.

There are basically three types of membranes for the kidney dialyzer (Fig. 10-12). The flat plate type was developed first and can have two or four layers. The blood passes through the spaces between the membrane layers while the dialysate is passed between the membrane and the restraining boards. The second and most widely used type is the coil membrane, in which two cellophane tubes (each 9 cm in circumference and

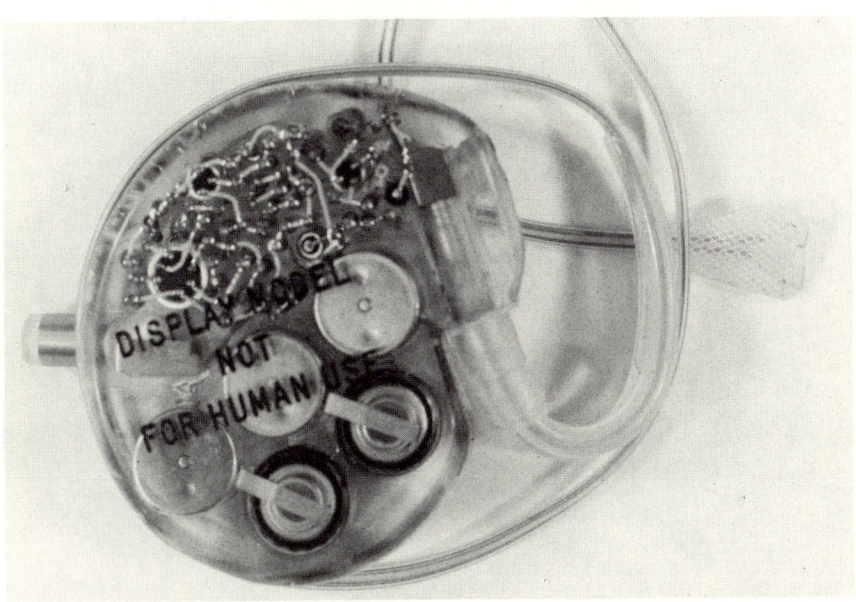

Figure 10-9. A typical pacemaker consists of a power source and electronic circuitry encased in a solid plastic. The electric wires are coated with a flexible polymer, usually silicone rubber.

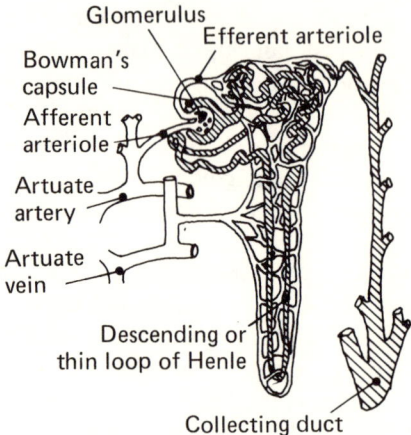

Figure 10-10. A kidney nephron.

108 cm long) are flattened and coiled with an open-mesh spacer material made of nylon. The newest addition is made of hollow fibers. Each fiber has dimensions of 225 and 285 μm inside and outside diameter and is 13.5 cm long. Each unit contains up to 11,000 hollow fibers. The blood flows through the fibers while the dialysate is passed through the outside of the fibers. The operational characteristics of the various dialyzers are given in Table 10-4.

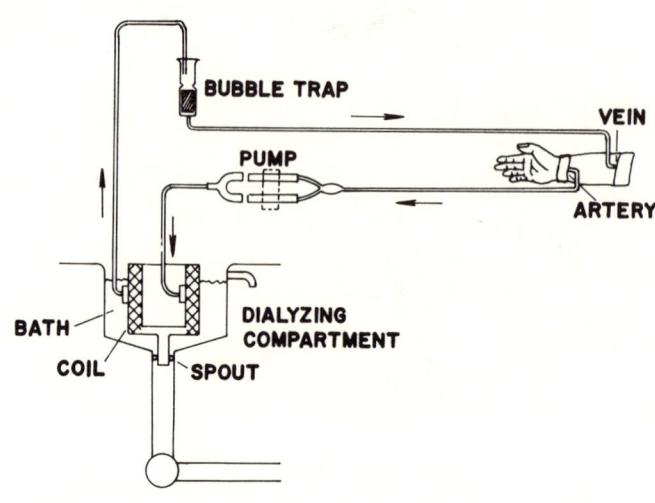

Figure 10-11. A typical dialyzer.

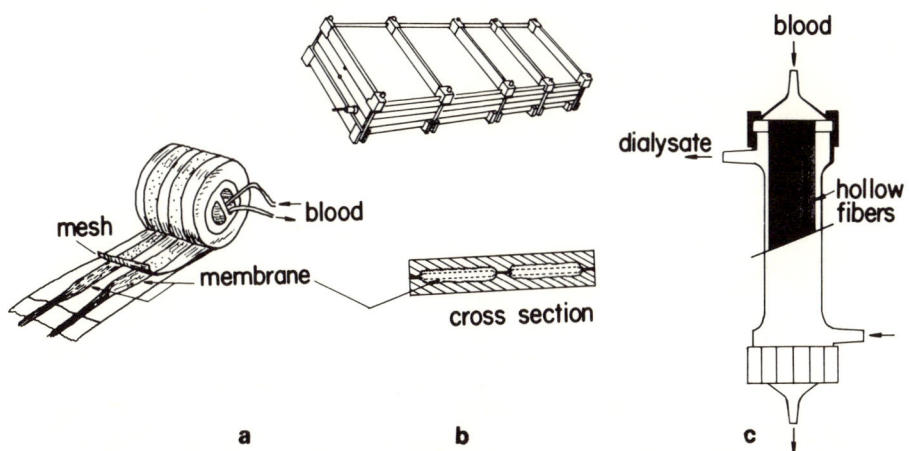

Figure 10-12. Three types of artificial kidney dialyzer: (a) twin coil, (b) flat plate, and (c) hollow fibers.

Recently there have been some efforts to improve dialyzers using charcoals. The blood can be circulated directly over the charcoal or the charcoal can be made into microcapsules incorporating enzymes or other drugs.

Most dialysis membranes are made from cellophane, which is derived from cellulose. Ideally, the membrane should selectively remove all the metabolic wastes as the normal kidney does. Specifically, the membrane should not selectively sequester materials from dialyzing fluid, should be blood compatible so that an anticoagulant is not needed, and should have sufficient wet strength to permit ultrafiltration without

Table 10-4. Comparison of the Plate and Coil Artificial Kidneys

Function	Flat plate (2 layers)	Coil (twin)
Membrane area (m^2)	1.15	1.9
Priming volume (ml)	130	1000
Pump needed?	No	Yes
Blood flow rate (ml/min)	140–200	200–300
Dialysate flow rate (liters/min)	2.0	20–30
Blood channel thickness (mm)	0.2	1.2
Treatment time (h)	6–8	6–8

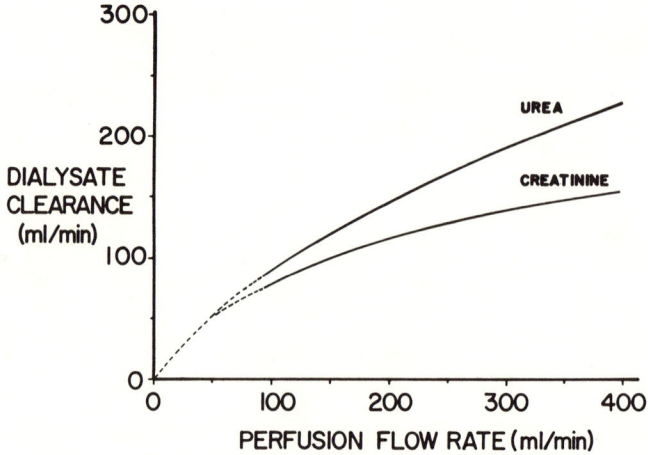

Figure 10-13. Average *in vitro* dialysance versus test solution flow rates. Fresh dialysis bath addition rate in a circulating single pass (RSP) dialyzer is 600 ml/min with bubble trap pressure of 60 mm Hg at 37°C. (Courtesy of Travenol Laboratories, Inc., Morton Grove, Illinois.)

significant dimensional changes. It should allow the passage of low molecular weight waste products while preventing passage of plasma proteins.

Two clinical grade cellophanes are available, Cuprophane® (Bemberg Co., Wuppertal, Germany) and Visking® (American Viscose Co., Fredericksburg, Va.). The cellophane films contain 25-Å-diameter pores which can filter molecules smaller than 4000 g/mol. A typical curve for flow rate versus clearance is shown in Figure 10-13. The clearance curve is flattened at high flow rate; this is the so-called square-meter concept, which is an important design factor in a membrane.

There have been many attempts to improve the cellophane membrane's wet strength by cross-linking, copolymerization, and reinforcement with other polymers such as nylon fibers. Also, like other blood-interfacing materials, the surface has been coated with heparin to prevent clotting. Other membranes, such as copolymers of polyethyleneglycol and polyethyleneterephthalate, can filter selectively due to their alternate hydrophilic and hydrophobic segments. Besides improving the membrane for better dialysis, the main thrust of kidney research is to make it more compact (portable or wearable kidney) and less costly (home dialysis, reusable or disposable filters, etc.).

The other important development in dialysis is the use of a cannula that must be connected to the blood vessels, as discussed in Chapter 9. To

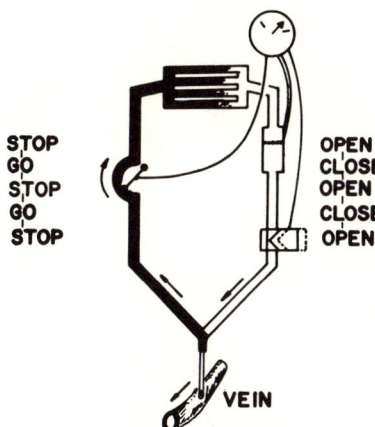

Figure 10-14. A schematic arrangement of the single-needle dialysis of Dr. Klaus Kopp. The pump operates continuously or intermittently synchronized with the inflow and outflow of blood. (Redrawn from W. J. Kolff, *Artificial Organs*, p. 151, J. Wiley and Sons, New York, 1976.)

minimize the repeated trauma on the blood vessels, the cannula is sometimes permanently implanted for chronic kidney patients. For the same reason, a single-needle dialysis technique has been developed, as shown in Figure 10-14.

Example 10-3

What will be the blood urea nitrogen concentrations after 5 and 10 hours of dialysis if the initial concentration is 1 g/liter? The concentration after dialysis can be expressed exponentially:

$$C_A = C_0 \exp\left(\frac{Q_B(\beta - 1)t}{V}\right)$$

where C_0 is the original dialysate concentration, Q_B is the blood flow rate, t is time, V is the volume of body fluid (60% of body weight), and β is a constant determined by the mass transfer coefficient (K), Q_B, and surface area (A) of the membrane by $\exp(KA/Q_B)$ (according to D. O. Cooney, *Biomedical Engineering Principles*, p. 332, Marcel Dekker, New York, 1976). The patient weighs 60 kg and a two-layer flat plate dialyzer was used.

Answer

$$\beta = \exp\left(-\frac{KA}{Q_B}\right)$$
$$= \exp\left(-\frac{1.15 \times 10^4 \text{ cm}^2}{63 \text{ min/cm} \cdot 170 \text{ ml/min}}\right)$$
$$= \underline{0.343}$$

For 5 h:

$$C_A = 1 \text{ (g/liter)} \exp\left(\frac{170 \text{ ml/min} (0.343 - 1) \times 5 \times 60 \text{ min}}{36{,}000 \text{ ml}}\right)$$
$$= \underline{0.39 \text{ g/liter}}$$

For 10 h:

$$C_A = 1 \text{ (g/liter)} \exp\left(\frac{170 \text{ ml/min} (0.343 - 1) \times 10 \times 60 \text{ min}}{36{,}000 \text{ ml}}\right)$$
$$= \underline{0.15 \text{ g/liter}}$$

Note that an extra 5 h of dialysis removed only 0.24 g/liter of urea nitrogen compared to the 0.61 g/liter removal of the first 5 h.

PROBLEMS

10-1. What characteristics of the cardiovascular system make using man-made materials in it different from implants in other parts of the body?

Answer
The major difference is that the cardiovascular system is in contact with blood. Another difference is the dynamic loading condition of arterial prosthesis.

10-2. Explain the principles of operation and the advantages and disadvantages of the single-needle dialysis machine (Figure 10-14) compared with a conventional machine.

Answer
The flow in the blood lines from the single needle to and from the patient is always unidirectional. Only the flow in the single needle is to and from the vein.

Obviously, single-needle dialysis is better for minimizing the trauma on the blood vessels. The pump can be used to operate the dialysate system; thus a wearable artificial kidney (WAK) can be made. There is some mixing of blood between dialyzed and undialyzed blood.

10-3. List two major technical problems associated with hemodialysis.

Answer
a. Damage on the formed elements of blood, such as RBC/WBC, platelets, etc.
b. Interfaces between the catheter and blood vessels.

10-4. List the advantages and disadvantages of a kidney transplant as compared with a dialysis machine.

SOFT TISSUE REPLACEMENT II

Answer
Advantages:

a. Complete freedom of movement away from the cumbersome dyalizer.
b. More complete restoration of kidney function especially secretion of enzymes, hormones, etc.
c. Less damage to the blood and no problem of infection, trauma, etc., due to the catheterization.
d. Psychologically better solution.
e. In the long run the cost may be lower.
f. Some patients cannot be machine-dialyzed.

Disadvantages:

a. Hard to obtain donors.
b. Problems related to rejection such as use of drugs; also, radiation can be detrimental to the patient.
c. The operation itself is major surgery and thus involves a certain danger.
d. Once rejected it is harder to have a second operation.

10-5. Calculate the tension developed on the wall of each fiber of the hollow fiber dialyzer by assuming that the maximum pressure differential of the inside and outside of the fiber is 10 mm Hg.

Answer
From equation (7-8) and $r \approx 130 \ \mu m$

$$T = P \cdot r$$
$$= 10 \text{ mm Hg} \cdot 133 \text{ Pa/mm Hg} \cdot 130 \times 10^{-6} \text{ m}$$
$$= 0.173 \text{ N/m}$$

Since the length of the fiber is 13.5 cm,

$$F = 0.173 \text{ N/m} \times 0.135 \text{ m}$$
$$= 0.0233 \text{ N}$$

Therefore, for a 30-μm wall thickness

$$\sigma = \frac{F}{A} = \frac{0.0233 \text{ N}}{30 \times 10^{-6} \text{ m} \times 0.135 \text{ m}} = \underline{5.8 \text{ kPa}}$$

10-6. From Figure 10-11 indicate areas in the system which are most likely to damage the blood flowing through it.

Answer
Arterial cannula, pump, coil, bubble trap, and the tube wall itself.

10-7. Calculate the molecular weight of the heparin monomer given in equation (10-1) of the text. If the average degree of polymerization is 30, what is the average molecular weight of the heparin?

Answer

$$C's: 24 \rightarrow 288$$
$$H's: 38 \rightarrow 38$$
$$O's: 20 \rightarrow 320$$
$$N's: 2 \rightarrow 28$$
$$678 \text{ g/mol}$$

$$M.W. = 674 \times 30 = \underline{20,320 \text{ g/mol}}$$

10-8. List major problems associated with pacemakers. One of the problems is the source of energy. Can a nuclear-powered energy source solve the problem?

Answer
Like any other implant the pacemaker has a problem of fixation of electrodes at the heart. The conduction of current decreases as the tissue heals and encapsulates the electrodes. The fatigue of lead wires and insulation of the wires and circuit sometimes present a problem.

The power source itself also limits the implant period (usually 5 years), although the nuclear-powered (plutonium 238 and promethium 147) pacemakers can prolong the source more than 30 years.

10-9. Describe methods of electrode attachments to the heart for the pacemaker. Can one make the surface of the electrode porous and let tissue grow into the interstices, thus making a viable interface between the electrode and the heart? What are the disadvantages and advantages of this system?

Answers
The myocardial electrodes are sewn into the muscle of the ventricles. In emergency cases the electrodes can be introduced transvenously and placed in the endocardium and paced externally. In experiments the porous electrodes at the tip have been shown to be feasible for fixation. However, if it continues to stimulate tissue ingrowth, the problem of threshold voltage elevation will be created, as with the conventional myocardial electrodes.

FURTHER READING

S. D. Bruck, *Blood Compatible Synthetic Polymers: An Introduction*, Charles C Thomas, Springfield, Ill., 1974.
D. O. Cooney, *Biomedical Engineering Principles*, Marcel Dekker, New York, 1976.
J. D. Hardy (ed.), *Human Organ Support and Replacement*, chapters 13, 16, and 17, Charles C Thomas, Springfield, Ill., 1971.
C. Homsy and C. D. Armeniades (eds.), *Biomaterials for Skeletal and Cardiovascular Applications*, Biomaterials Materials Symposium No. 3, J. Wiley and Sons, New York, 1972.

R. L. Kronenthal and Z. Oser (eds.), *Polymers in Medicine and Surgery,* Plenum Press, New York, 1975.

H. Lee and K. Neville, *Handbook of Biomedical Plastics,* chapters 3 and 5, Pasadena Technology Press, Pasadena, Calif., 1971.

R. I. Leininger, "Polymers as Surgical Implants," *CRC Crit. Rev. Bioeng., 2,* 333, 1972.

G. H. Myers and V. Parsonnet, *Engineering in the Heart and Blood Vessels,* J. Wiley and Sons, New York, 1969.

L. Vroman and F. Leonard (eds.), "The Behavior of Blood and Its Components at Interfaces," *Ann. N.Y. Acad. Sci., 283,* 1976.

Transactions, American Society for Artificial Internal Organs (published yearly; contains studies related to this chapter).

CHAPTER 11

HARD TISSUE REPLACEMENT I: LONG BONE REPAIR

Hard tissues can be divided into bones and teeth, as mentioned previously. This chapter will be concerned with repairs or replacements of bone with implants. Joint and tooth implants will be studied separately in Chapter 12.

It is logical that bone repairs should be made according to the best repair route that the natural tissues follow, i.e., if they heal faster when compressive force or strain is exerted then compression should be provided through an appropriate implant design. On the other hand, if compression is detrimental for wound healing, the opposite approach should be taken. Unfortunately, the effect of compressive or tensile force on long bone repair is not fully understood. Worse yet, experimental results show that a completely opposite conclusion can be arrived at.

The secret of osteogenic and osteoclastic activity is believed to be related to the normal activities of the bone *in vivo*. Thus, the equilibrium between osteogenic and osteoclastic activity can be balanced according to the static and dynamic force applied *in vivo*; i.e., if more load is applied, the equilibrium tilts toward more osteogenic activity to counteract the load, and vice versa. Of course this should be done without excessive load, which can damage the cells rather than enhance their activity.

This cause and effect may also be related to the piezoelectric phenomenon of bone and other tissues in which the compressive strain on the tissue can induce electric potentials that may trigger the tilting of the equilibrium. This is the basis of the electrically stimulated fracture repair of clinical nonunions.

Although the mechanics and design of implants are outside the scope of this book, some information on these subjects may help in selecting the best material for a particular application.

The design principles and criteria for orthopedic implants are the same as for any other engineering applications that require a dynamic load-bearing member. Although it is tempting to duplicate the natural tissues with materials having the same strength and shape, this has not been practical because the natural tissues have one major advantage over the man-made materials, i.e., their ability to adjust to a new set of circumstances by remodeling their micro- and macrostructures. Consequently, the fatigue of tissues is minimal unless disease hinders the natural processes.

The first axiom of design in engineering is to make the product simple. This will not only minimize the chance of complication, such as infection during and after implantation, but also reduces deterioration of the implants by fatigue and corrosion as discussed previously (Chapters 2 and 4). Other factors to be considered in the actual design of an implant are rigidity against bending, e.g., bone plates and femoral neck nails; rigidity and strength, e.g., spinal rod; and strength against tensile and compressive fracture.

11.1. INTERNAL FRACTURE FIXATION DEVICES

11.1.1. Wires and Screws

Although the exact mechanism of bone fracture repair is not yet known, one factor of fracture fixation is clearly important, i.e., stability of the implant with respect to the wound surfaces. Whether the fixation is accomplished by compressive or tensile force, the wound should be rigidly fixed so that the healing processes cannot be disturbed by unnecessary micro- or macromovement. This fixation can be accomplished by devices with a variety of shapes and sizes, as shown in Figure 11-1.

The simplest implants are the various metal wires (Kirschner wires if the diameter is less than $3/32$ inch or 2.38 mm and Steinman pins for the larger diameter) that can be used to hold fragments of bones together. Another simple device is the screw, which can be used alone or in combination with the onlay fracture plate. Screws are usually self-tapping for better fixation in the bone. The design of the screws to be used with bone plates should consider (1) ease of insertion through both the plate and

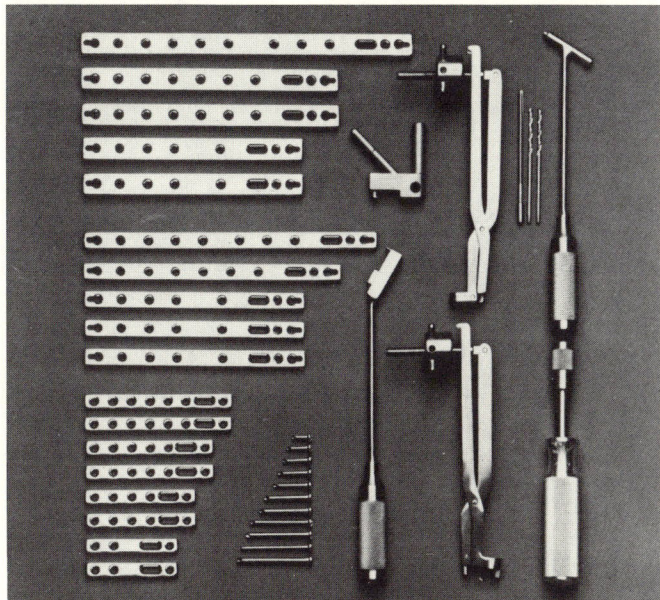

Figure 11-1. Various sizes of fracture plates and tools for insertion. (Courtesy of Zimmer, USA, Warsaw, Ind.)

bone, (2) fatigue strength to resist dynamic loading, and (3) firm attachment combined with ease of removal.

Table 11-1 illustrates the strength of various screws to pull out from the cortical bone and their original tensile strength. The safety factors (ratios between the tensile strength of screw and anchorage) in these cases are less than 3, which is the normal engineering applications for this type

Table 11-1. Strength of Bone Screws[a]

Screw	Outside diameter (mm)	Root diameter (mm)	Mean strength of anchorage in cortical bone (N)	Mean tensile force for fracture (N)	Safety factor
Sherman	3.7	2.7	1746	4011	2.3
AO type	4.5	3.1	2275	5012	2.2
Experimental type	4.8	4.0	2354	6963	2.96

[a] From O. Lindahl, *Acta Orthopaed. Scand.*, *38*, 101, 1967, by permission of the publisher.

Table 11-2. Greatest Resistable Bending Moment at Proximal End of Femur[a]

Muscle group	Number of subjects tested		Bending moment (newton meters)	
	Men	Women	Range	Mean
Hamstring	11		54–93	72
		17	26–54	35
Quadriceps	6		42–60	51
Hip abductors	6		38–108	63
		3	24–48	39
Hip adductors	6		60–126	81
		3	32–40	30

[a] From M. Laurence, M. A. R. Freeman, and S. A. V. Swanson, *J. Bone Joint Surg.*, *51B*, 754, 1969, by permission from the publisher.

of dynamic loading situation. A discussion of screws is not complete without mentioning the head. Although it is not critical what types of screw heads are used, the grooves should be sufficiently deep to prevent slipping of the screwdriver when inserting or removing the screws. The presence of fats and tissue fluids coupled with poor visibility make this simple task very difficult.

11.1.2. Fracture Plates

There are many different types and sizes of fracture plates, as shown in Figure 11-1. Because the forces generated by the muscles in the limbs are very large (see Table 11-2) the plates must be strong. This is especially

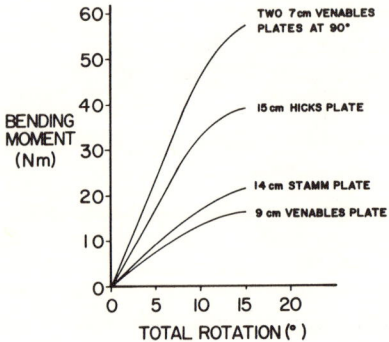

Figure 11-2. Bending moment versus total rotation of various bone plates. (Redrawn after M. Laurence, M. A. R. Freeman, and S. A. V. Swanson, *J. Bone Joint Surg.*, *51B*, 754, 1969, by permission from the publisher.)

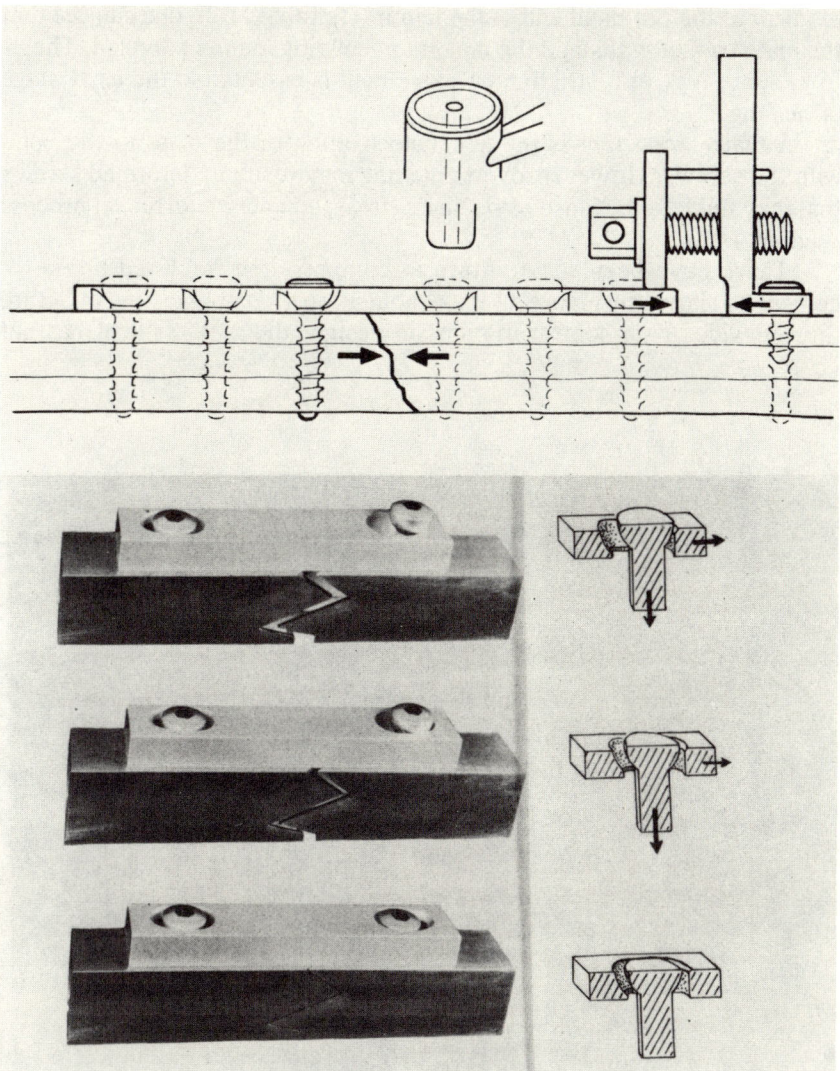

Figure 11-3. Principle of a dynamic compression plate (DCP): method of compression with a device (upper) and the principle (lower) (from M. Allgöwer, P. Matter, S. M. Perren, and T. Rüedi, *The Dynamic Compression Plate, DCP*, p. 18, Springer-Verlag, New York, 1973, by permission from the publisher.)

true for the femoral and tibial plates. The bending moment and angle of various devices are plotted in Figure 11-2. In comparison with the bending moment at the proximal end of the femur (cf. Table 11-2), one can see that the plate cannot withstand the maximum bending moment applied. Therefore, some type of restriction on movement is essential in the early stage of healing.

Equally important is the adequate fixation of the plate to the bone with the screws. However, overtightening may result in deformed screws that may fail later because of strain or stress, an energy corrosion process (section 4.3).

There have been some efforts to compress the fractured bones together, as shown in Figure 11-3. Although most companies manufacture these devices with some variations in design, the new concept has not

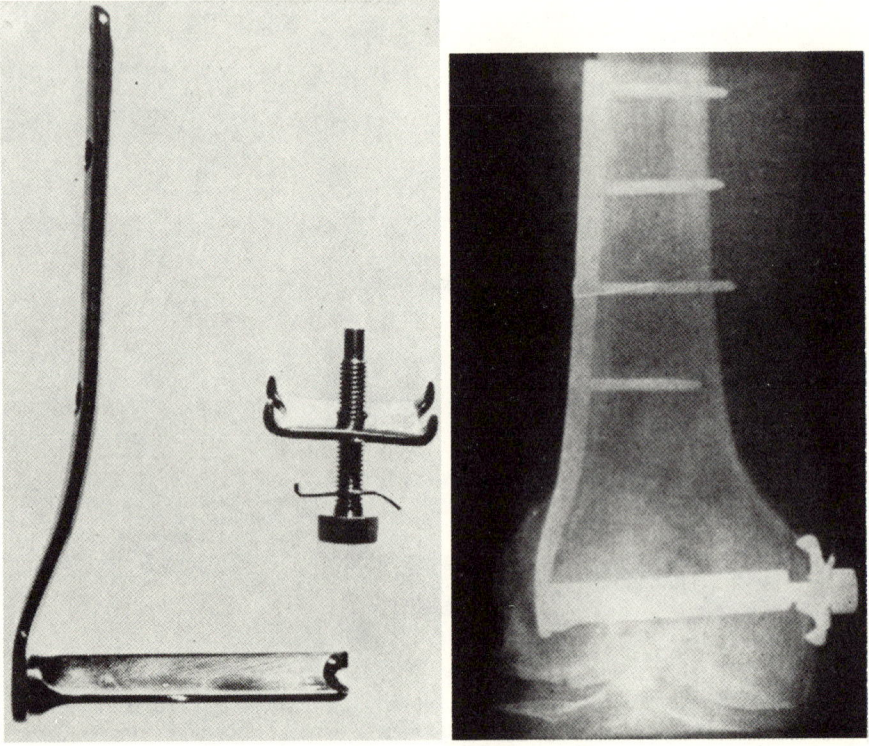

Figure 11-4. Devices to fix a cancellous bone of a supracondylar fracture of the femur. (A. Brown and J. C. D'Arcy, *J. Bone Joint Surg.*, *53B*, 420, 1971, by permission from the publisher.)

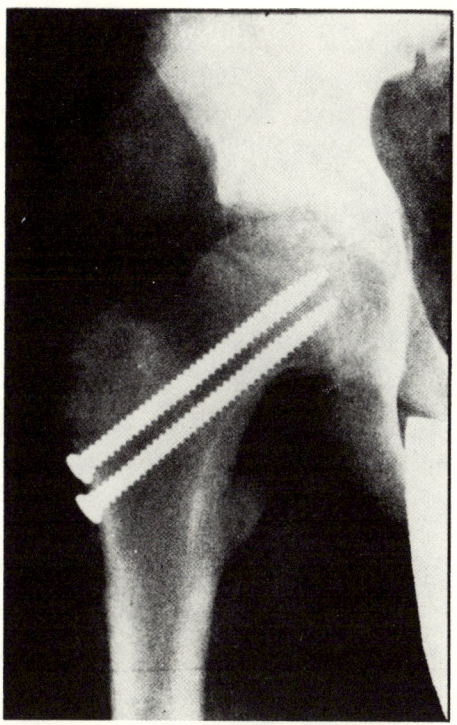

Figure 11-5. An example of a simple fracture fixation of a cancellous bone. (Reproduced by permission from H. M. Frost, *Orthopaedic Biomechanics,* p. 444, Charles C Thomas, Springfield, Ill., 1973.)

been used extensively mainly because of the added complexity of the devices and controversy as to whether the compressive force or strain is beneficial.

Considerable care must be exercised when fixing cancellous bone because there are far fewer bone speculae to support the load than in cortical bone. An example of the fixation of the ends of a long bone is shown in Figure 11-4, in which the fractured bones are fixed with a combination of screws, plates, bolts, and nuts. The bulk necessary for adequate stabilization of the fracture increases the chance of infection near the site of implants.

Sometimes a cancellous bone fracture can be fixed by using a simple nail, as shown in Figure 11-5. This is a special case because the patient was a young child (who incidentally has 2–3 times the trabecullar bone mass in the cancellous femoral head and neck region); and because the epiphyseal plate lies close to the hip joint, the loading is essentially normal to the fracture surface. Also the freely mobile hip joint relieves stress except during the compression cycle which forces the parts together. Obviously, a wide range of choices is available. The choice is largely determined by the surgeon, not by the patient or bioengineer.

11.1.3. Intramedullary Devices

Intramedullary devices are used to fix fractures of the long bones by inserting them snugly into the intramedullary cavity. Sometimes a stabilizer is used with this device. This type of implant should have some spring in it to exert some elastic force inside the bone cavity to prevent rotation and to fix the fracture firmly.

This close fitting of intramedullary nails is sometimes difficult if the bone is curved in the middle. Also, at the point of contact, the constant compressive pressure may result in bone resorption. It is also obvious that

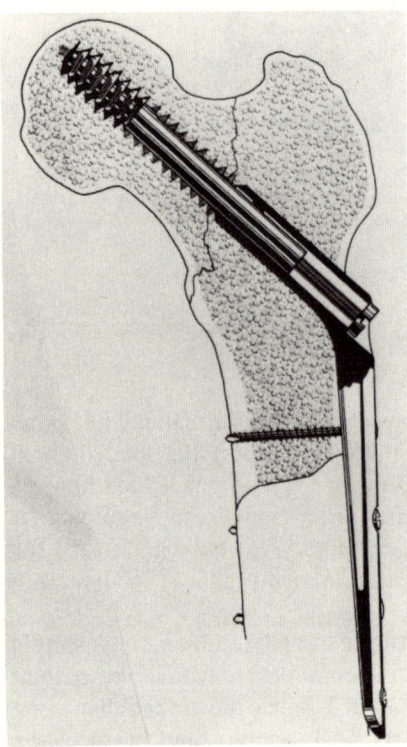

Figure 11-6. An example of hip nail fix of a fractured femoral head. (Courtesy of DePuy, Division of Bio-Dynamics, Inc., Warsaw, Ind.)

Example 11-1

Measure the angle of the nail and plate in Figure 11-6. Calculate the maximum static bending moment at the neck of the implant if the body weight (60 kg) is applied at a 20° angle in the vertical position.

Answer

The maximum bending moment about the neck of the implant is $W_2 \cdot L_o \cos 40°$, where $W_2 = W_o \cos 20°$; hence the maximum bending moment is $\underline{W_o L_o \cos 40° \cos 20°}$.

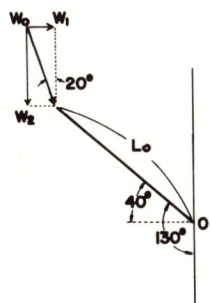

If a person weighs 60 kg and the L_o is 15 cm, the maximum moment is 60 kg · 9.8 m/s² · 15 × 10⁻² m cos 40° cos 20° = $\underline{63.5 \text{ N} \cdot \text{m}}$ (cf. Table 11-2).

11.1.4. Nail and Plate Devices for Femoral Osteotomy

Although occasionally the fractured femoral head can only be fixed with nails or screws, it is more common to use nails (or screws) and a plate at the same time (Fig. 11-6). This is usually true in intertrochantric osteotomy because the simple nail and plate cannot absorb the dynamic load exerted on the device. The various designs for this purpose are shown in Figure 11-7. The insertion of these devices is rather difficult, although a set of special tools and guide wires are supplied by the manufacturers.

Hip nails have several designs, as shown in Figure 11-8. All but one have holes in the middle to guide accurate insertion. All the designs have tried to minimize the amount of material while maximizing the bending moment (M), which was derived in chapter 2 [equation (2-31)]:

$$M = EI/\rho \qquad (11\text{-}1)$$

where E is the modulus of elasticity and I is the moment of inertia. Therefore, the stiffness of the material is important as well as the moment of inertia, which in turn is related to the section modulus, $Z (= I/c)$. The various cross sections of the hip nails of Figure 11-8 are designed to

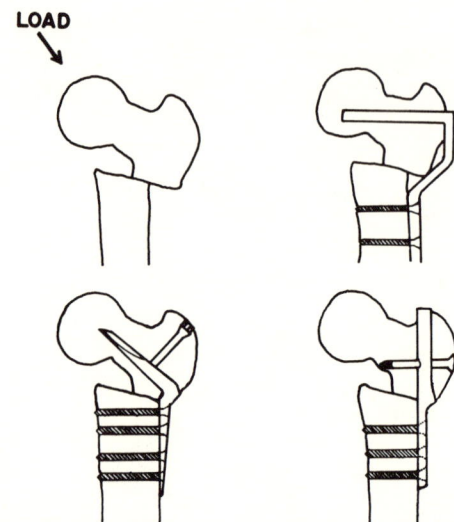

Figure 11-7. Schematic diagram of osteotomy devices: (*a*) instable state after femoral osteotomy; (*b*) typical osteotomy device; (*c*) Wainwright-Hammond osteotomy plate; (*d*) Osborne-Ball osteotomy plate. (By permission from D. F. Williams and R. Roaf, *Implants in Surgery,* pp. 407, 451; W. B. Saunders Co., London, 1973.)

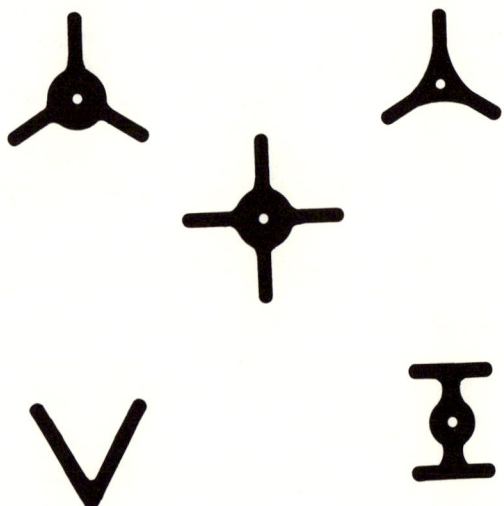

Figure 11-8. Cross section of various hip nails. (By permission from D. F. Williams and R. Roaf, *Implants in Surgery,* p. 402, W. B. Saunders Co., London, 1973.)

increase the bending moment through increased section modulus with a minimum amount of material.

This economic section can be demonstrated by comparing the amount of material needed for the same section modulus of solid and hollow cylinders (see Fig. 2-12 for section modulus):

$$\left.\frac{\pi r^3}{4}\right|_{solid} = \left.\frac{\pi(R^4 - r^4)}{4R}\right|_{tube} \tag{11-2}$$

Equation (11-2) can be reduced to

$$\left(\frac{R}{r}\right)^4 - \frac{R}{r} - 1 = 0 \tag{11-3}$$

The approximate solution to equation (11-3) gives $R = 1.22r$. Therefore, for the same section modulus, the outer radius of the hollow tube should be about 22% larger than the inner radius. This translates into about one-half the cross-sectional area required for the hollow cylinder than for the solid cylinder. This is why the hollow cylinder is more economical in resisting bending.

Example 11-2

Assuming identical shapes, list, in the order of resistance to bending, bone plates made of the following substances:

a. Al_2O_3
b. Type 316 cold-worked stainless steel
c. Compact bone (femur)
d. Polymethylmethacrylate
e. Wrought Co–Cr alloy

Answers

The order of the most resistance to bending can be made according to the modulus of elasticity because bending stress is proportional to it.

e. Wrought Co–Cr alloy	230 GPa
b. Type 316 cold-worked S.S.	200 GPa
a. Al_2O_3	24 GPa
c. Compact bone	17.2 GPa
d. Polymethylmethacrylate	3 GPa

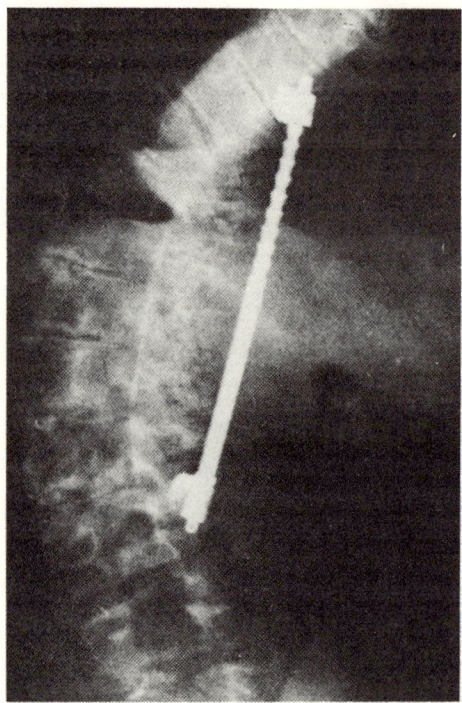

Figure 11-9. Harrington spinal distraction rod. (Reproduced by permission from D. F. Williams and R. Roaf, *Implants in Surgery,* p. 456, W. B. Saunders Co., London, 1973).

11.1.5. Spinal Fixation Devices

When the spinous elements are deformed in such a manner that the anterior elements are longer than the posterior ones, the resulting structure is bent backward; this condition is called lordosis. The opposite condition is called kyphosis. In severe cases of such spinal deformities, an internal and external fixation is used for correction. Several designs exist to stabilize or straighten the curvatures, one of which is shown in Figure 11-9. Other designs include plates that are attached to the spinous processes by using bolts and expanding spinal jacks that are hooked to the spine through the articular process so that the distraction can be adjusted during implantation.

The main problems with these devices are (1) adjusting or extending the device as the spine is straightened and (2) necrosis of the bones where the fixation device is attached. Necrosis results from the tremendous moment exerted by the trunk muscles—more than 100 newton meters.

HARD TISSUE REPLACEMENT I

As the spine is straightened, it is harder to distract without hooks because the leverage on the spine decreases. Multiple hooks are sometimes attached to obviate this problem.

Example 11-3

To correct spinal luxation a surgeon is experimenting with a bone cement and Steinmann pins on dogs. A bioengineer, trying to test the strength of the union by making test samples, obtained a load deflection curve for the original sample *in vitro* and then cemented the same vertebrae and tested as shown in the figure.

a. Calculate the maximum bending shear stresses if the cross section of the spinal disk can be assumed to be a circle with a diameter of 1.4 cm, and the distance between the point of loading and the disk is 5.3 cm.
b. If the maximum deflection is 0.5 cm, how long it will last? Assume that the maximum bending shear stress can be used as the tensile stress of Figure 6-4.

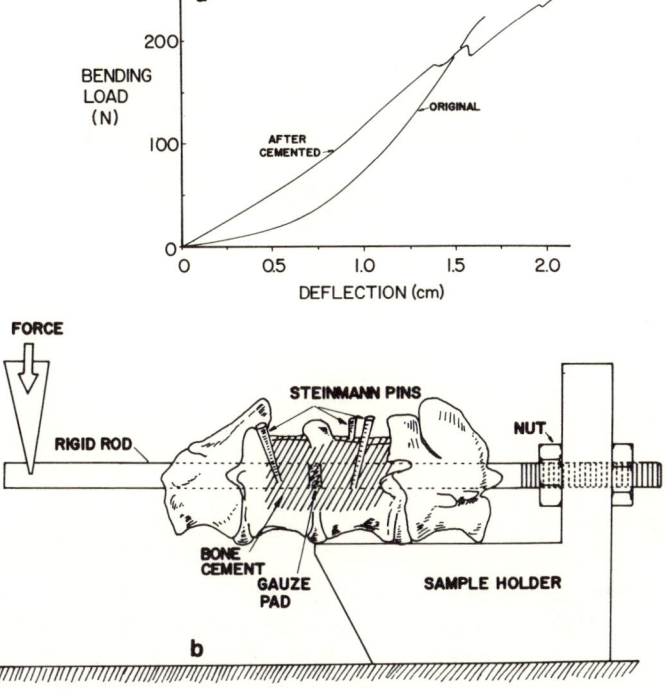

Example 11-3 figure. (a) Load-deflection curves for spinal fixation study. Figure (b) shows how the tests were made. (Courtesy of G. Rouse, J. B. Park, G. H. Kenner, and C. L. Gendreau.)

Answers

a. The maximum shear stress can be expressed as

$$\sigma_{max} = M/Z$$

where Z is the section modulus and M is the bending moment. For a circle $Z = \pi r^3/4$ (Figure 2-12) and $M = P \times a$ (load × distance); hence

$$\sigma_{max} = 4P \cdot a/\pi r^3$$

For the original sample,

$$\sigma_{max} = \frac{4 \cdot 230 \text{ N} \cdot 5.3 \text{ cm}}{\pi (0.7 \text{ cm})^3}$$
$$= \underline{45.3 \text{ MPa}}$$

For the cemented sample,

$$\sigma_{max} = \frac{4 \cdot 250 \cdot 53}{\pi (0.7 \text{ cm})^3}$$
$$= \underline{49.2 \text{ MPa}}$$

b. At 0.5-cm deflection, the load on the cemented sample is about 50 N;

$$\sigma = \frac{4 \cdot 50 \text{ N} \cdot 5.3 \text{ cm}}{\pi (0.7 \text{ cm})^3}$$
$$= \underline{9.8 \text{ MPa}}$$

Compared with Figure 6-4 it will last about 10^5 cycles.

11.2. MATERIALS USED FOR INTERNAL FRACTURE FIXATION DEVICES

The materials currently used for internal fracture fixation devices are all made of metals and their alloys. Several alloys, such as stainless steels, titanium, and cobalt chromium alloys, have enjoyed wide acceptance because they can be easily obtained in high quality and have excellent resistance to corrosion *in vivo*. This resistance has been the most important aspect of metallic devices because of their electrochemical activity in solution in the body.

11.2.1. Stainless Steels

The most common stainless steel used for orthopedic implant work is AISI type 316; its composition is given in Table 11-3. The most important structural feature of any stainless steel is its maintenance of the austenitic γ phase, as shown in Figure 4-5. This can be achieved by controlling the relative amount of alloying elements. Figure 11-10 shows that for nickel and

Table 11-3. Specifications for Stainless Steel, Type 316, for Use in Implant Surgery

Element	Weight percent (w/o)
Chromium	17.00–20.00
Nickel	10.00–14.00
Molybdenum	2.00–4.00
Carbon	0.08 max
Silicon	0.75 max
Manganese	2.00 max
Sulfur	0.03 max
Phosphorus	0.03 max
Iron	Balance

[a] From *Annual Book of ASTM Standards,* part 46, p. 377, American Society for Testing and Materials, Philadelphia, 1976, courtesy of ASTM.

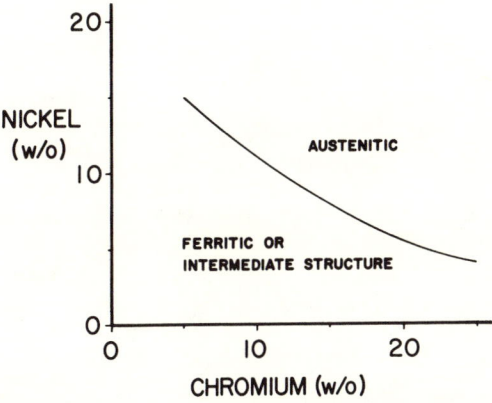

Figure 11-10. The partial phase diagram of stainless steel (0.1–0.5 w/o carbon) in relation to the nickel and chromium content. (Adapted from J. H. G. Monypenny, *Stainless Iron and Steel,* vol. 2, p. 115, Chapman & Hall, London 1954.)

chromium, the γ phase can be formed 18 w/o Cr, 8 w/o Ni, and 0.1 w/o C but not for, e.g., 18 w/o Cr, 5 w/o Ni, and 0.1 w/o C.

By adding molybdenum to stainless steels to enhance corrosion resistance the α phase (ferrite) may be stabilized to such an extent that the steel may consist of two phases, i.e., α and γ. To avoid this, the nickel content of molybdenum-bearing stainless steels must be increased. However, this increase causes more intergranular corrosion than with a lower nickel content molybdenum steel. Lowering the carbon content or adding other stabilizing agents such as titanium and niobium prevents such corrosion.

This discussion indicates why the composition given in Table 11-3 for stainless steel is the most desirable. Table 11-4 shows mechanical properties of some stainless steels. Because the austenitic phase does not have decreasing solid solubility, which is prerequisite for the precipitation hardening (Section 4.2), the usual way of hardening is cold working. As noted in Table 11-4, cold-working does not increase the modulus of elasticity, although it makes a substantial increase in yield strength and a marginal increase in fracture strength. As mentioned in Chapter 4, the trade-off is decreased fracture strain.

11.2.2. Cobalt–Chromium Alloys

The cobalt–chromium alloys commercially called Stellite®, Vitallium®, Vinertia®, etc., vary somewhat in composition. The alloy can be manufactured into final products by either a cast or a wrought process. The different compositions are given in Table 11-5. The small

Table 11-4. Mechanical Properties of Some Stainless Steels[a]

Property	18 Chromium, 8 nickel fully softened	Extra-low carbon, 18 chromium, 10 nickel fully softened	Type 316 fully softened	Type 316 cold-worked
Yield strength (MPa)	200–230	200–250	240–300	700–800
Tensile strength (MPa)	540–700	540–620	600–700	1,000
Ductility (elongation %)	50–65	55–60	35–55	7–10
Young's modulus (GPa)	200	200	200	200
Hardness (V.P.N.)	175–200	170–200	170–200	300–350
Fatigue limit (MPa)	230–250	—	260–280	300

[a] From D. F. Williams and R. Roaf, *Implants in Surgery*, p. 305, W. B. Saunders Co., London, 1973, by permission from the publisher.

Table 11-5. The Composition of Cobalt–Chromium Alloys[a]

	Cast alloys	Wrought alloys
Chromium	27.0–30.0[b]	19.0–21.0
Molybdenum	5.0–7.0	—
Nickel	2.5	9.0–11.0
Tungsten	—	14.0–16.0
Carbon	max	0.05–0.15
Silicon	1.0	1.0
Manganese	1.0	2.0
Iron	0.75 max	3.0 max
Cobalt	Balance	Balance

[a] From *Annual Book of ASTM Standards*, part 46, pp. 382, 385; American Society for Testing and Materials, Philadelphia, 1976, courtesy of ASTM.
[b] Figures in weight percent.

amount of molybdenum in the casting alloys decreases the grain size during solidification, thus increasing the strength (see section 4.2). The alloy is readily work-hardened as a result of its inability to deform plastically. Therefore, an implant produced by cold-working has to be annealed. The mechanical properties of the cobalt–chrome alloys are slightly better than the stainless steels, as shown in Table 11-6. Furthermore, they are excellent in both corrosion and fatigue resistance.

The cast alloys are frequently used in dental work for their excellent reproducibility of details. The casting procedure for this alloy is the "lost

Table 11-6. Mechanical Properties of Cobalt–Chromium Alloys[a]

Property	Cast alloy	Wrought alloy (solution-annealed)	Wrought alloy (cold-worked)
Yield strength (MPa)	450	380	1050
Tensile strength (MPa)	655	900	1540
Ductility (elongation %)	8	60	9
Young's modulus (GPa)	200	230	230
Hardness (V.P.N.)	300	240	450

[a] From D. F. Williams and R. Roaf, *Implants in Surgery*, p. 310, W. B. Saunders Co., London, 1973, and *Annual Book of ASTM Standards*, part 46, pp. 382, 385; American Society for Testing and Materials, Philadelphia, 1976.

wax" investment technique. In this procedure the grain size and its distribution are controlled by controlling solidification temperature, mold surface, and the presence of other elements (e.g., molybdenum) to achieve a homogeneous solid solution that results in superior physical properties.

Example 11-4

Calculate the densities of 316 stainless steel and cast Co-Cr alloy based on their composition.

Answer

Since the density of an alloy can be expressed by

$$\rho = \rho_1 W_1 + \rho_2 W_2 + \cdots + \rho_n W_n$$

where the subscripts designate each component of the alloy and W is the weight fraction. Therefore, from Tables 11-3 and 11-5,

$$\rho_{s.s.} = 7.19 \times 0.185 \text{ (Cr)} + 8.9 \times 0.12 \text{ (Ni)} + 10.2 \times 0.03 \text{ (Mo)}$$
$$+ 7.43 \times 0.02 \text{ (Mn)} + 7.87 \times 0.64 \text{ (Fe)}$$
$$= \underline{7.89 \text{ g/cm}^3}$$

$$\rho_{Co-Cr} = 7.19 \times 0.285 \text{ (Cr)} + 10.2 \times 0.06 \text{ (Mo)} + 8.9 \times 0.025 \text{ (Ni)} + 2.4$$
$$\times 0.01 \text{ (Si)} + 7.43 \times 0.01 \text{ (Mn)} + 7.87 \times 0.0075 \text{ (Fe)}$$
$$= \underline{8.38 \text{ g/cm}^3}$$

11.2.3. Other Metals

Titanium and its alloys are used as implant materials because of their high corrosion resistance and relatively low density (4.5 g/cm³ compared with 7.9, 8.3, and 9.2 g/cm³ for stainless steel and cast and wrought cobalt–chrome alloys, respectively). The excellent corrosion resistance of pure titanium results from the tenacious oxide film (TiO_2) on the surface, as in aluminum oxide (Al_2O_3), which protects the surface from further oxidation. Unlike aluminum, the titanium oxide film is very stable in saline solution at room temperature.

Because of the difficulty of obtaining titanium in pure form and of machining, it is alloyed with other elements such as aluminum, vanadium, manganese, silicon, molybdenum, and tin. The mechanical properties of titanium and its alloys are given in Table 11-7.

Another corrosion-resistant metal used for implant manufacture is

Table 11-7. Mechanical Properties of Surgical Titanium and Its Alloys[a]

Type	Tensile strength (MPa)	Yield strength (0.2% offset, MPa)	Elongation (%)
Grade 3, pure	415[b]	345	18
Grade 4, pure	550	485	15
Ti-6Al-4V	896	830	10

[a] From *Annual Book of ASTM Standards,* part 46, pp. 381, 398; American Society for Testing and Materials, Philadelphia, 1976, courtesy of ASTM.
[b] All values are minimum requirements.

tantalum. It forms an oxide film like that of titanium. Its mechanical properties are similar to that of stainless steel but tantalum has a high density, 16.6 g/cm^3; however, the difficulty of extracting it from ore makes the metal less attractive.

Platinum metals (platinum, ruthenium, rhodium, palladium, osmium, and irridium) and their alloys have been utilized from time to time for surgical implants and devices, although their high cost combined with relatively poor mechanical properties restricts their use to such special cases as dental bridges and electrodes for special circumstances (e.g., pacemaker tips).

PROBLEMS

11-1. Indicate where the metallic corrosion will predominate in the osteotomy devices shown in Figure 11-7.

Answer
The interfaces between the screws and stem, sharp corners in the stem.

11-2. Explain why the wrought cobalt–chromium alloy does not contain molybdenum and the cast alloy does.

Answer
The molybdenum increases the strength of the cast Co–Cr alloy; however, cold-working increases the strength of the wrought Co–Cr alloy and hence the latter does not need molybdenum. (The modulus of elasticity of molybdenum is 345 GPa, about 70% higher value than the Co–Cr alloy.)

11-3. Write the weight percentages of three major alloying elements for 18/8, 18/10, and type 316 stainless steel alloys and wrought Co–Cr alloys. Determine from Figure 11-10 if they are austenitic or ferritic.

Answer

Element	18/8	18/10	316	Wrought Co–Cr
Cr	18	18	17–20	19–21
Ni	8	10	10–14	9–11
Co	—	—	—	45.85–51.95
Fe	>50	>50	59–68	—

From Figure 11-10, all are <u>austenitic</u>.

11-4. Calculate the tensile stresses of the three types of bone screws listed in Table 11-1. Compare with the values given in Tables 11-4 and 11-6. Can you tell what possible kinds of metals each is made from?

Answer

Types of screw	Fracture strength (MPa)	Possible metals
Sherman	700	Co–Cr or cold-worked 316 s.s.
AO type	664	Softened s.s.
Experimental	554	18/8 or 18/10 s.s.

11-5. Explain the principle of a dynamic compression plate (Fig. 11-3). What problems may arise by using this device? Will this device be more susceptible to corrosion than the conventional bone plate–screw system?

Answer

Figure 11-3 is self-explanatory. The degree of compression is hard to determine and some fractures cannot be compressed by this technique. Because the screws and plate are fixed in a strained state they are more susceptible to corrosion than in a strain-free system.

11-6. Calculate the tensile strength and density ratios for the following materials:

a. Type 316 cold-worked stainless steel
b. Cast Co–Cr alloy
c. Ti–6Al–4V alloy
d. High-density polyethylene
e. Polymethylmethacrylate
f. Pyrolytic carbon (compressive strength)
g. Al_2O_3 crystals (compressive strength)

Answers

a. $1{,}000/7.9 = 127$ MPa cm³/g
b. $655/8.3 = 79$ MPa cm³/g
c. $896/4.5 = 199$ MPa cm³/g
d. $30/0.96 = 31$ MPa cm³/g
e. $70/1.20 = 58$ MPa cm³/g
f. $517/1.75 = 295$ MPa cm³/g
g. $862/3.8 = 227$ MPa cm³/g

FURTHER READING

C. O. Bechtol, A. B. Ferguson, and P. G. Laing, *Metals and Engineering in Bone and Joint Surgery,* Ballière, Tindall, and Cox, London, 1959.

J. H. Dumbleton and J. Black, *An Introducton to Orthopedic Materials,* chapter 9, Charles C Thomas, Springfield, Ill. 1975.

V. H. Frankel and A. H. Burstein, *Orthopedic Biomechanics,* chapter 7, Lea & Febiger, Philadelphia, 1971.

G. Kuntscher, *The Practice of Intramedullary Nailing,* Charles C Thomas, Springfield, Ill., 1947.

C. S. Venable and W. C. Stuck, *The Internal Fixation of Fractures,* Charles C Thomas, Springfield, Ill., 1947.

D. F. Williams and R. Roaf, *Implants in Surgery,* chapters 6–8, W. B. Saunders Co., London, 1973.

CHAPTER 12

HARD TISSUE REPLACEMENT II: JOINTS AND TEETH

The articulation of joints poses more problems than long bone fracture repairs. These include wear, corrosion, and their products, as well as complicated dynamics of movement. In addition, the massive nature of totally replacing such joints as the knee and elbow near the skin also renders the possibility of infection greater. More important, if the replacement fails for any reason it is harder to replace the joint a second time because a large portion of the natural tissue has been destroyed.

For these reasons orthopedic surgeons try to salvage the existing joint and use implants as a last resort. However, the hip prosthesis has gained large support in recent years for older patients.

Such familiar tooth implants as amalgams for cavities are outside the scope of this book. Instead total tooth replacement with man-made materials will be considered.

Tooth replacement is challenged by the transcutaneous (or percutaneous) nature of the severe oral environment, which continually changes its chemical composition, pH, temperature, etc. Teeth undergo the most severe compressive stress in the body (up to 850 N) and a satisfactory material or technique has not yet been found that can withstand not only the compressive stress but also the added torque and shear stress during mastication.

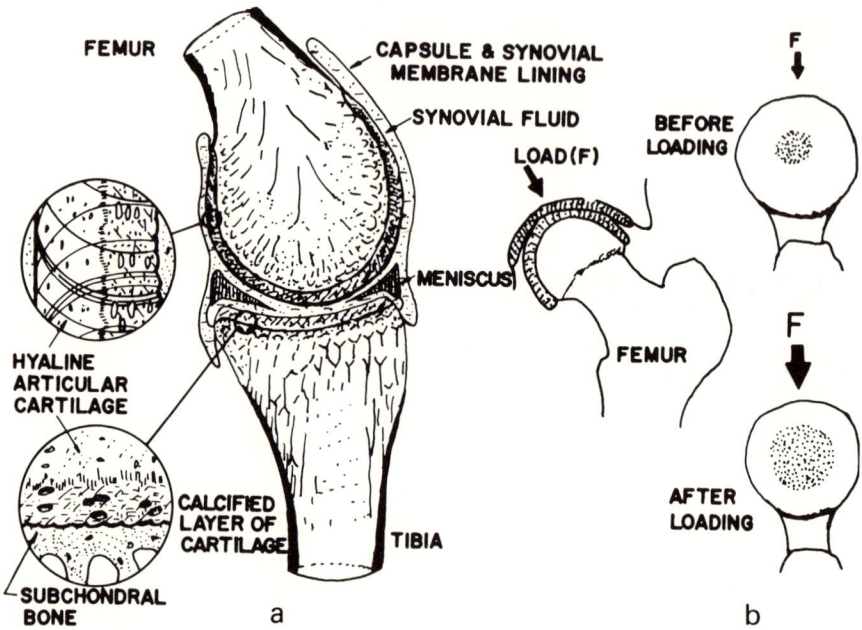

Figure 12-1. Structural arrangement of a knee joint seen from the side (a) and hip joint before and after loading (b). (From H. M. Frost, *Orthopaedic Biomechanics,* pp. 269, 277; Charles C Thomas, Springfield, Ill., 1973, by permission from the publisher.)

12.1. STRUCTURE AND FUNCTION OF JOINTS

The hip and shoulder joint have ball-and-socket articulation, while other joints such as knee and elbow are of the hinge type. However, all joints possess two opposing smooth cartilaginous articular surfaces that are lubricated by viscous synovial fluid. The fluid is made of polysaccharides (section 7.1.2) that adhere to the cartilage and on loading can be permeated out onto the surface to reduce friction. The cartilage is not vascularized, and the repair and nutrition of the tissues appear to be diffusional processes.

Nature provides the surface of the joint with a large area to minimize load concentration effect (Fig. 12-1) for the hip and knee joint. The loading can be absorbed further by the trabecular subchondral bone underlying the cartilaginous tissue; it also transfers the load gradually.

The actual articulation of the joint is performed by the ligaments, tendons, and muscles. An example of this anatomic aspect is given in Figure 12-2. The analysis of forces acting on the various tendons and ligaments is very complicated. Even the center of rotation of the knee joint cannot be determined with any great precision; in fact, it shifts position with each movement. The eccentric joint movement helps to distribute the load throughout the entire joint surfaces.

Some joints, such as the knee, have fibrous, cartilaginous minisci shaped like wedges located between the sliding surfaces (Figure 12-1). The main function of the miniscus is believed to transfer the load over a larger area than is possible without it.

The joint forces applied during a range of activities are given in Table 12-1. Of course, the forces applied during walking vary considerably with each motion, as shown in Figure 12-3. It should not be surprising that these forces are up to 8 times the body weight, because they act on the joints in a dynamic rather than a static manner. This type of biomechanical analysis can help in designing a better implant.

12.2. VARIOUS JOINT REPLACEMENTS

Traditionally joints have been restructured or remodeled by a surgical technique called arthroplasty. Resection and interposition ar-

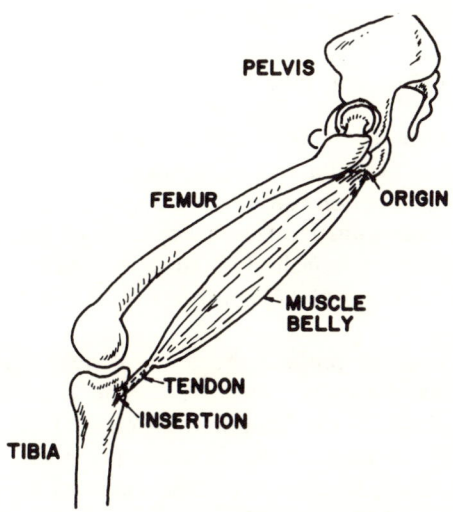

Figure 12-2. Arrangement of loading elements in a thigh. Note the attachment of muscle. (From H. M. Frost, *Orthopaedic Biomechanics,* p. 262, Charles C Thomas, Springfield, Ill., 1973, by permission from the publisher.)

Table 12-1. Average Maximum Values of Forces at Hip and Tibiofemoral Joints during a Range of Activities[a]

Activity	Maximum joint force (multiples of body weight)	
	Hip	Knee
Level walking		
Slow	4.9	2.7
Normal	4.9	2.8
Fast	7.6	4.3
Up stairs	7.2	4.4
Down stairs	7.1	4.9
Up ramp	5.9	3.7
Down ramp	5.1	4.4

[a] From J. P. Paul, "Loading on Normal Hip and Knee Joints and Joint Replacement," in *Advances in Hip and Knee Joint Technology,* ed. M. Schaldach and D. Hohmann, p. 63, Springer-Verlag, Berlin, 1976.

throplasty are common methods of surgical operation. The replacement arthroplasty involves the use of an implant.

12.2.1. Hip Joint Replacement

The early methods of correcting diseased or fractured hip joints involved only the acetabular cup or femoral head. One technique of restoring the hip joint function is to place a cup over the femoral head while the surface of acetabulum is also resected to fit the cup. The implant serves as a mold interposing the two surfaces which eventually readjust according to the function of the joint. An example of this technique is illustrated in Figure 12-4.

Some have tried to replace the femoral head after resection with various designs, as shown in Figure 12-5. The wide variety of over 30 different implants reflects the limited knowledge of the function of joints and the ability of the joint to take almost any insult imposed on it by various implants. Most femoral head replacements are done with the installation of an acetabular cup. This is the so-called total hip joint replacement, which is frequently performed bilaterally. The various types of hip implants can be grouped into ball and socket, retained ball and socket, trunnion bearing, floating acetabulum, and double cup (Fig. 12-6).

The single most difficult problem in hip joint as well as other joint replacements is the fixation of the implants, because the implant lies on cancellous bone, which has few trabecullae to support the large load imposed. Also, the stress concentration of the implant at points of sharp contact, such as the calcar region and the end of the femoral stem, makes the already weakened bone more necrotic. In fact, the first wide acceptance of total hip replacement was achieved by providing an acceptable fixation using acrylic bone cement. The cement is inserted when it becomes doughy, after mixing the polymer powder and monomer liquid thoroughly, and the prosthesis is press-fitted into the drilled hole. The cement can also be used in the fixation of the acetabular cup.

The cement not only serves as the initial attachment of the implant with bone, but also acts as a shock absorber because it is a viscoelastic polymer. This seemingly small difference in the rigidity among the cement,

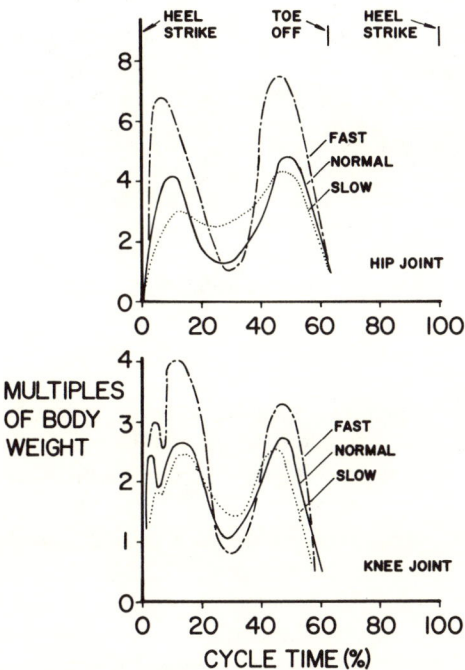

Figure 12-3. Variation of forces with time of hip and knee joint in walking. (J. P. Paul, "Loading on Normal Hip and Knee Joints and Joint Replacements," in *Advances in Hip and Knee Joint Technology,* ed. M. Schaldach and D. Hohmann, Springer-Verlag, Berlin, 1976, by permission from the publisher.)

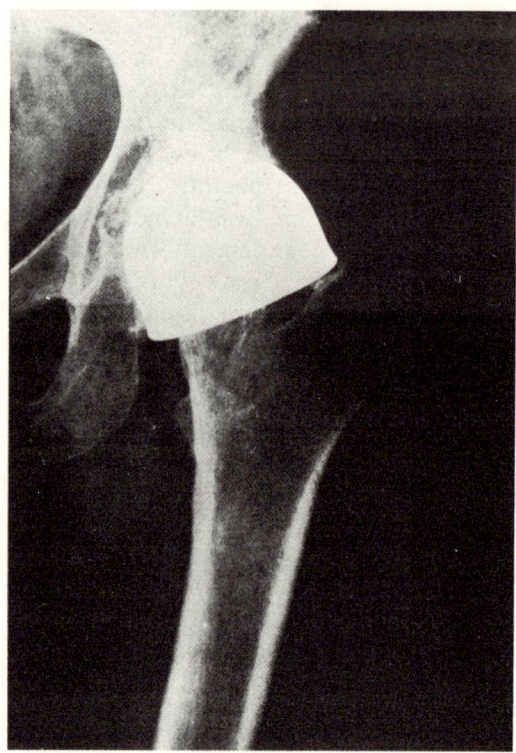

Figure 12-4. An example of mold arthroplasty. (From H. M. Frost, *Orthopaedic Biomechanics*, p. 299, Charles C Thomas, Springfield, Ill., 1973, by permission from the publisher.)

bone, and prosthesis helps to spread the load over a large area and to obviate the stress concentration problem. However, the stress on the bone in the distal stem is much higher than in the proximal region (calcar) when the stem is inserted by using bone cement, as shown in Figure 12-7. This causes bone resorption of the calcar region, which in turn will lead to either loosening or fracture of the stem.

To obviate this problem a higher loading condition in the proximal region is desirable by making the neck portion of the stem longer. However, this arrangement increases the moment applied in the midstem, causing fracture more readily. Actually, the new trends in stem design and insertion technique are to make it straighter, thus decreasing the moment.

The cement itself sometimes gives problems, e.g., the monomer interferes with the systemic function and decreases the blood pressures.

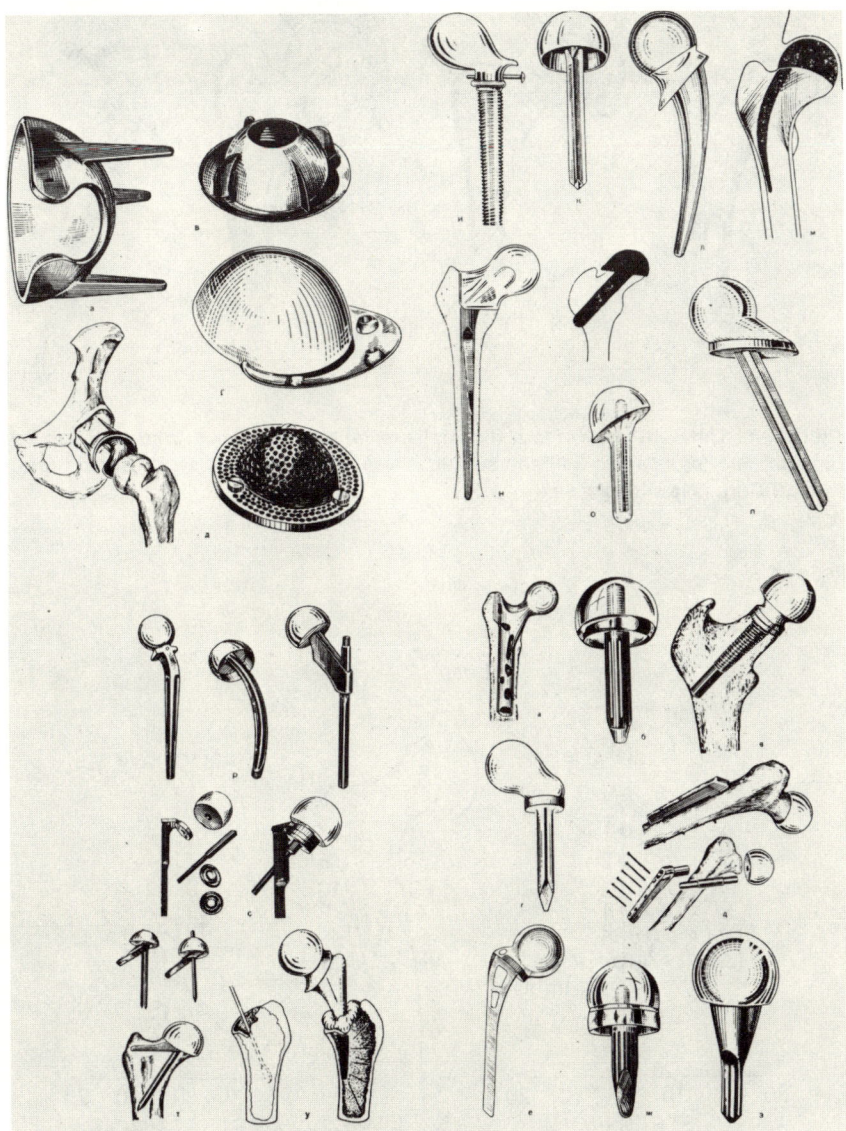

Figure 12-5. Various designs of acetabular and femoral head components of hip prostheses. (From K. M. Sivash, *Alloplasty of the Hip Joint, A Laboratory and Clinical Study*, Medical Press, Moscow, 1967, by permission.)

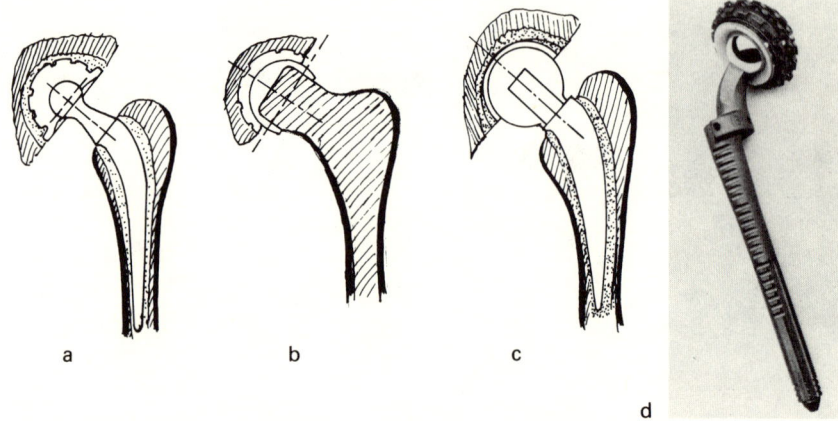

Figure 12-6. Different types of total hip implants: (*a*) ball and socket, (*b*) double cup, (*c*) trunnion, and (*d*) retained ball and socket. (Sivash design, courtesy of United States Surgical Corp., New York.)

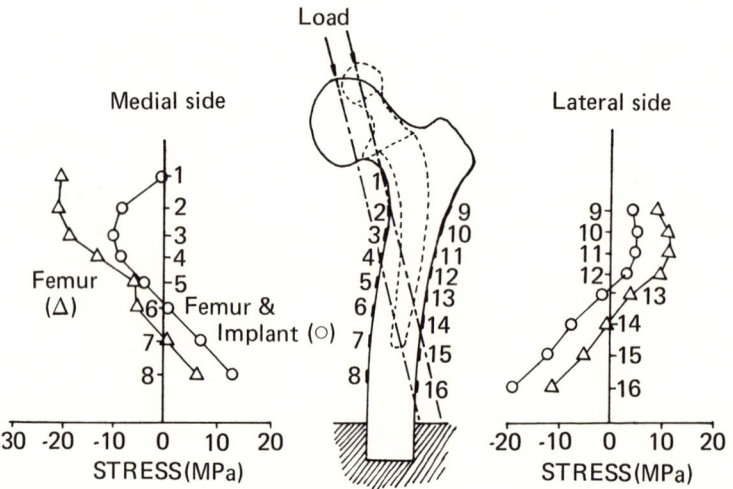

Figure 12-7. The stresses on the surface of the femoral stem by a load of 4000 N. The numbers indicate the location of strain gauges to measure the deformations. Note that there is no stress in position 1 (calcar region) after insertion of the implant. (From S. A. V. Swanson and M. A. R. Freeman, *The Scientific Basis of Joint Replacement*, p. 41, J. Wiley and Sons, New York, 1977.)

The highly exothermic polymerization reaction can cause a local temperature of over 60°C, which can result in cell necrosis. Also the extensive intramedullary cavity preparation for the cement space can block the bone sinusoids, enhancing tissue necrosis and fat embolism.

Another added problem is the difficulty and the extent of tissue destroyed when removing implants for any reason. The replaceability of the implant is an important aspect of its design. In this regard, the Sivash implant has an inherent weakness in its design, i.e., the whole prosthesis has to be replaced even if only one component has failed. Also the fixation of the device is accomplished by direct apposition to the bone, which results in stress concentration at the end of the stem and sharp edges of the acetabular cup in contact with bone.

The friction between the ball and cup of the hip joint is significant when it creates a rotational torque. Especially at high loading rates the frictional torque becomes very significant for the cobalt–chromium alloy hip joint, as shown in Figure 12-8. The stainless steel/polyethylene and cobalt–chromium alloy/polyethylene combinations are better for reducing frictional torque and wear than the all-metal system. The high frictional torque of the all-metal system may also be due to the larger surface contact because the femoral head is much larger than the metal–polymer prosthesis. In actual use the all-metal system works well without exerting high frictional torque because tissue fluids lubricate the surfaces.

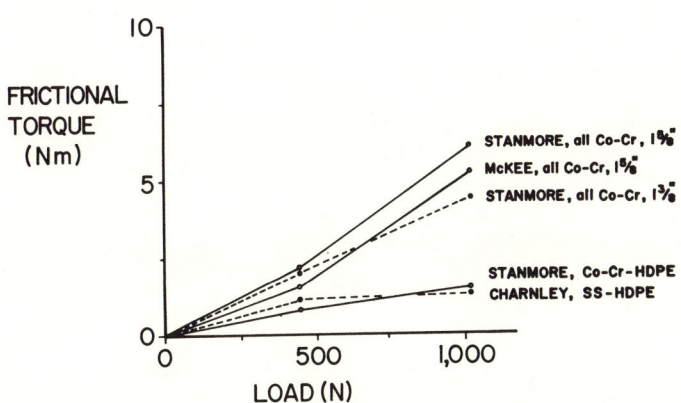

Figure 12-8. Frictional torque versus applied load for various hip prostheses. (Redrawn from J. N. Wilson and J. T. Scales, *Clin. Orthopaed. Related Res.*, 72, 145, 1970.)

Example 12-1

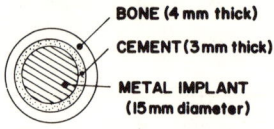

A bioengineer is trying to determine the amount of gap developed between bone and cement when a femoral hip is placed. The system is assumed to consist of concentric cylinders. Calculate the gap developed between bone and cement if the temperatures of cement, implant, and bone reach 55, 50, and 45°C, respectively, throughout each component uniformly (assume that the α value of the implant is $17 \times 10^{-6}/°C$).

Answer

From Table 2-1 the linear coefficients of thermal expansion are 8.3 and $81.0 \times 10^{-6}/°C$ for bone and cement; therefore, the shrinkage after equilibration with body temperature for each component will be

implant: $\Delta l = 1.5 \text{ cm} \times 17 \times 10^{-6}/°C(37-50)°C = -3.32 \ \mu m$

cement: $\Delta l = 0.6 \text{ cm} \times 81.0 \times 10^{-6}/°C(37-55)°C = -8.75 \ \mu m$

bone: $\Delta l = 0.8 \text{ cm} \times 8.3 \times 10^{-6}/°C(37-45)°C = -0.53 \ \mu m$

Since the metal implant shrinks, the diameter is only 3.32 μm, but the cement cannot shrink the full 8.75 μm because of the stiff implant. Therefore, the real gap between bone and cement is ½(3.32–0.53)μm, which is 1.4 μm. The cement will impose a hoop stress around the implant. If the hoop stress becomes large enough, the cement may break by its own shrinkage stress. (Cement breakage and the micro gap between the implant and bone have been observed in a clinical situation, although the former is a rare case.) The linear shrinkage of

$$0.145\% (8.75 \ \mu m/0.6 \text{ cm} \times 100)$$

is close to the reported value for the cement (0.12–0.27% for Simplex-P), although the α used is for solid glassy polymethalmethacrylate.

Example 12-2

The most widely used bone cement in hip joint replacement is made of powder and liquid, as given below (Simplex-P Bone Cement, Howmedica, Inc., Rutherford, N.J.).

HARD TISSUE REPLACEMENT II

Liquid

Methylmethacrylate monomer	97.4 v/o
N,N-dimethyl-p-toluidine	2.6 v/o
Hydroquinone	75 ± 15 ppm

Powder

Polymethylmethacrylate	16.7 w/o
Methylmethacrylate-styrene copolymer	83.3 w/o

(Radiopaque cement has 10 w/o barium sulfate in powders)

a. Write the polymerization process(es).
b. Why are N,N-dimethyl-p-toluidine and hydroquinone added in the liquid?
c. What would be the reason behind using the MMA-styrene copolymer?
d. Suggest ways to sterilize the cement.

Answers

a. The polymerization is free-radical-type (Section 6.1.2); initiation is achieved by peroxides present in the powder.

$$\underset{\text{MMA}}{\begin{array}{c} H \\ | \\ C \\ | \\ H \end{array} = \begin{array}{c} CH_3 \\ | \\ C \\ | \\ COOCH_3 \end{array}} \longrightarrow -CH_2-\underset{COOCH_3}{\overset{CH_3}{C}}-CH_2-\underset{COOCH_3}{\overset{CH_3}{C}}-CH_2-\underset{COOCH_3}{\overset{CH_3}{C}}-\cdots$$

$$\underset{\text{Styrene}}{\begin{array}{c} H \\ | \\ C \\ | \\ H \end{array} = \begin{array}{c} CH \\ | \\ C \\ | \\ \text{Ph} \end{array}} \longrightarrow -CH_2-\underset{COOCH_3}{\overset{CH_3}{C}}-CH_2-CH(\text{Ph})-CH_2-\underset{COOCH_3}{\overset{CH_3}{C}}-\cdots$$

b. The N,N-dimethyl-p-toluidine is added to accelerate the decomposition of peroxides, thus achieving rapid polymerization (hence it is called an *accelerator*). The hydroquinone is added to prevent the polymerization of the monomer (thus it is called an *inhibitor*).

c. Although the mechanical properties of the copolymer are not much different from those of the PMMA, the chemical properties are enhanced by the copolymerization.

d. The powders are sterilized by γ-radiation and the liquid is sterilized by the membrane filtration method. (Many other commercial bone cements are available, although Simplex-P is the only cement allowed to be used as a drug by FDA at this time).

12.2.2. Other Joint Replacements

The development of prostheses for joints other than the hip has been relatively slow and they are still not widely accepted by the medical profession. The requirements for a successful knee joint implant are the same as for other joints, including hip joint replacements. These include a low frictional torque without sacrificing the range of motion, a low wear rate, a rigid and viable fixation of prosthesis to the host, and replaceability.

The knee prosthesis has several severe inherent problems, i.e., loosening and infection because of its location. Some prostheses (e.g., McKeever) use the principles of mold arthroplasty of hip joint replacement, while others use the hinge principle, as shown in Figure 12-9.

Other joint replacements (elbow, shoulder, and ankle) have been given little attention compared to the hip and knee joints. The mechanical and design problems associated with these joints are the same as those of hip and knee joints. Some commercially available replacements for these joints are shown in Figure 12-10.

The finger joint prosthesis has been popularized by the immediate improvements of cosmetic appearance as well as functionality for severe arthritic joints. There are essentially two types of prosthesis—the integral hinge prosthesis and the mechanical hinge prosthesis (Fig. 12-11). The integral hinge type is made of silicone rubber or polypropylene. Sometimes silicone rubber is reinforced with woven fabrics. The success of this type of prosthesis lies on the fact that the stem is gliding through the intramedullary cavity encapsulated with a thin collagenous tissue. This is the same principle as the mold prosthesis discussed previously. The hinge type of prosthesis sometimes has two pronged stems to prevent rotation. (The Schulz finger prosthesis stems shown in Fig. 12-11 have a rectangular shape to prevent rotation.)

12.2.3. Materials Used for Joint Replacements

The materials used for joint replacements are the same materials studied so far. As mentioned in the hip prosthesis section, because friction and wear are major problems for articulating joints the materials that yield the least friction and wear have been chosen to construct the prosthesis. Polytetrafluoroethylene (Teflon®) was originally thought to be a good candidate because of its low friction, but it failed badly because of unfavorable wear properties. Interestingly, the acrylic polymer used for bone cement was used in the early design of the femoral head (Judet). Needless to say, the implants deteriorated fast and excessive wear resulted.

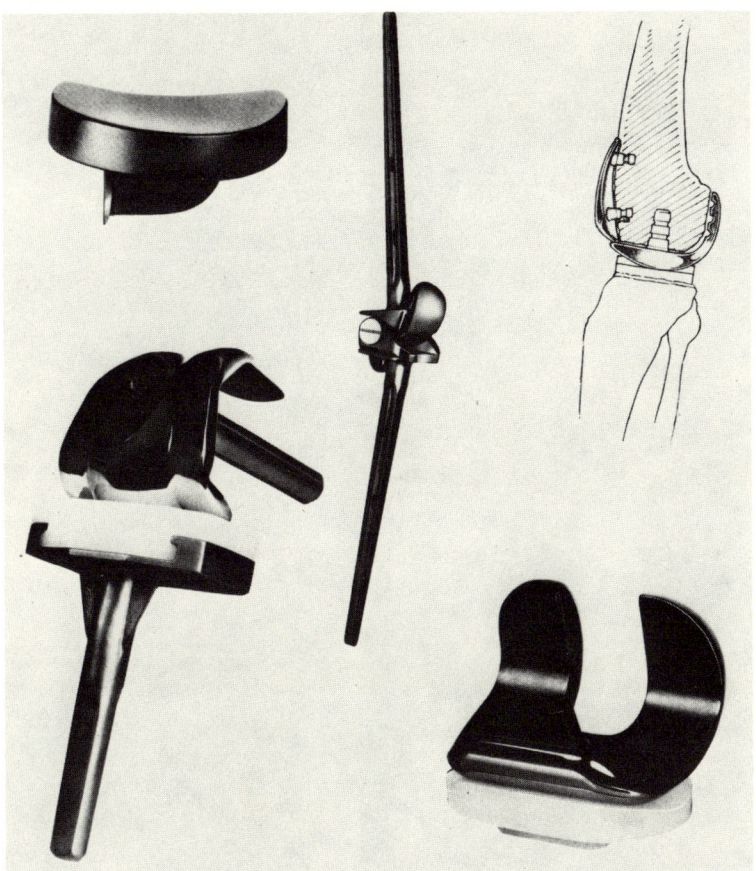

Figure 12-9. Various knee implant designs. Top left: McKeever tibial plateau. Bottom left: Variable axis total knee. Center Waldius total knee. Top right: patello-femoral components in position. Bottom right: Freeman–Swanson total knee. [All are from Howmedica, Inc., Rutherford, N.J., except (*d*), Richards Mfg. Co., Memphis, Tenn.]

As mentioned in conjunction with hip prosthesis, the Charnley hip prosthesis uses a high-density polyethylene (sometimes called ultrahigh molecular weight polyethylene) acetabular cup with cobalt–chromium alloy or stainless steel for the femoral component. This combination gives the least friction (cf. Fig. 12-8) and wear. Furthermore, the particles of polyethylene show a minimal tissue irritation, another important factor for this type of implant. Because the polyethylene acetabular cup is translucent to X rays it is sometimes circumferentially banded with a wire. Some prostheses are designed to replace the polyethylene compo-

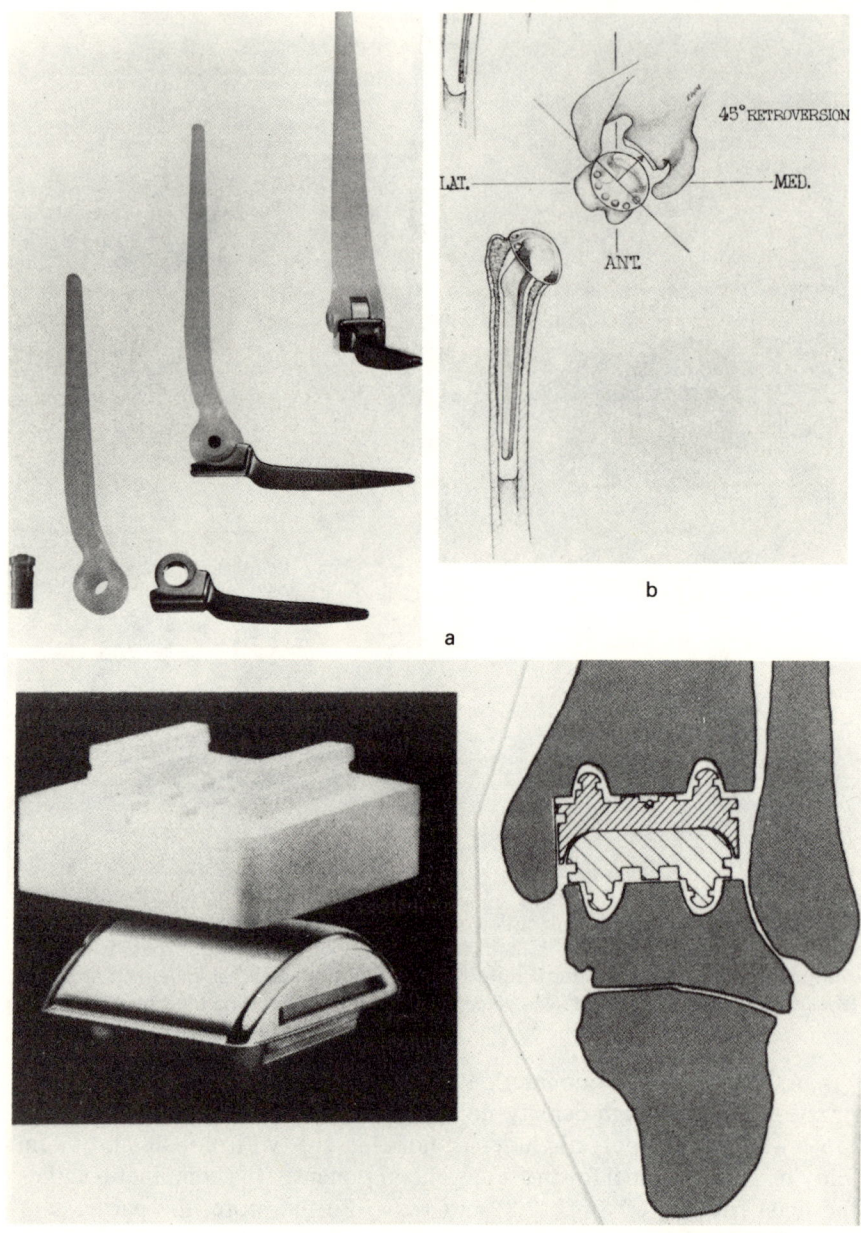

Figure 12-10. Typical prosthesis of (a) an elbow (Pritchard-Walker design, DePuy, Division of Bio-Dynamics, Inc., Warsaw, Ind.) (b) shoulder (Bechtol design, Richards Mfg. Co., Memphis, Tenn.) and (c) ankle (Mayo total ankle, DePuy).

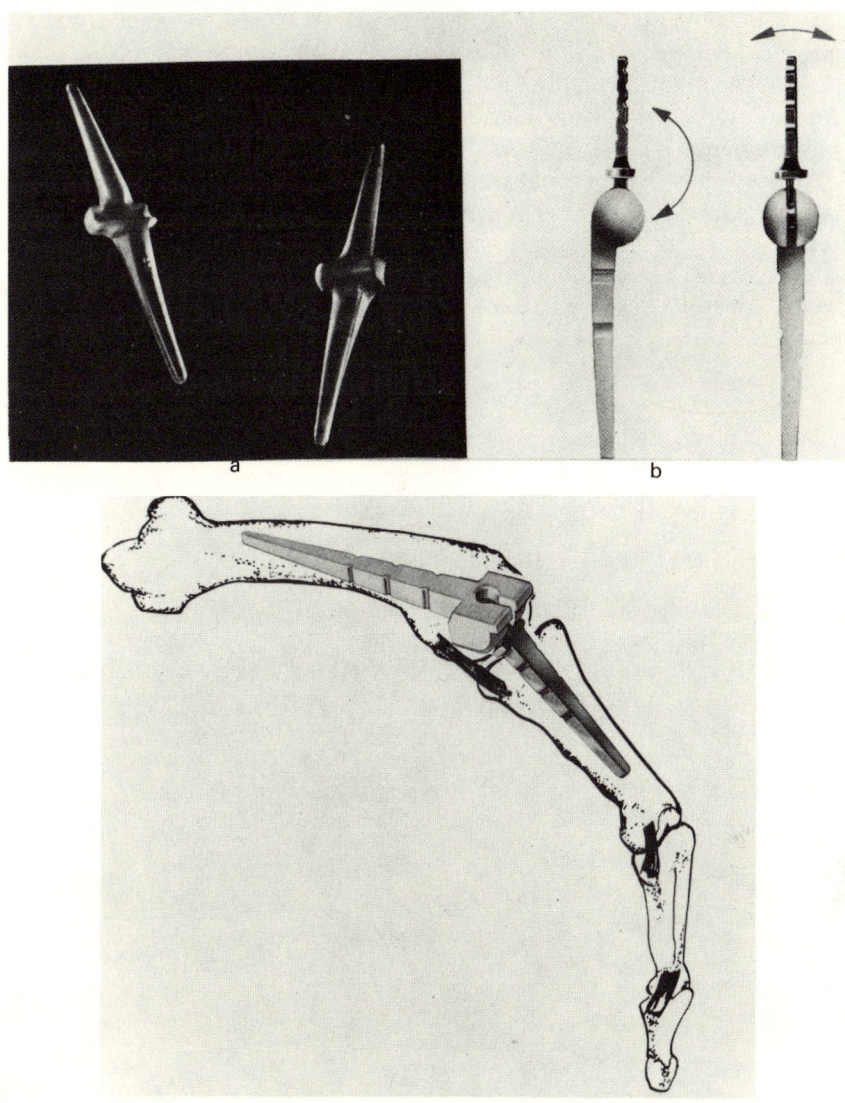

Figure 12-11. Finger joint prostheses: (a) is an integral hinge type; (b) and (c) are mechanical hinge type. (a, Swanson design, Dow Corning Co.; b, St. Georg design, Richards Mfg. Co.; c, Schulz design, Zimmer, USA.)

nent (either cup or femoral head) easily because the polymer wears out faster than metal, as shown in Figure 12-12. Some prostheses use different alloys in different parts. For example, the stem and acetabular components of the Sivash hip prostheses (cf. Fig. 12-6d) are made of titanium (Ti-6Al-4V) alloy, while the ball is made of cast cobalt–chromium alloy. Furthermore, the socket liner is made of an ultrahigh-molecular-weight polyethylene.

There has been a great deal of interest in developing wear-resistant materials for joint prostheses even though the polymer-metal combination seems to be satisfactory. Especially in France and Germany, dense alumina (Al_2O_3) has been studied clinically for use in joint replacements. The main advantages of the ceramic materials are their inertness

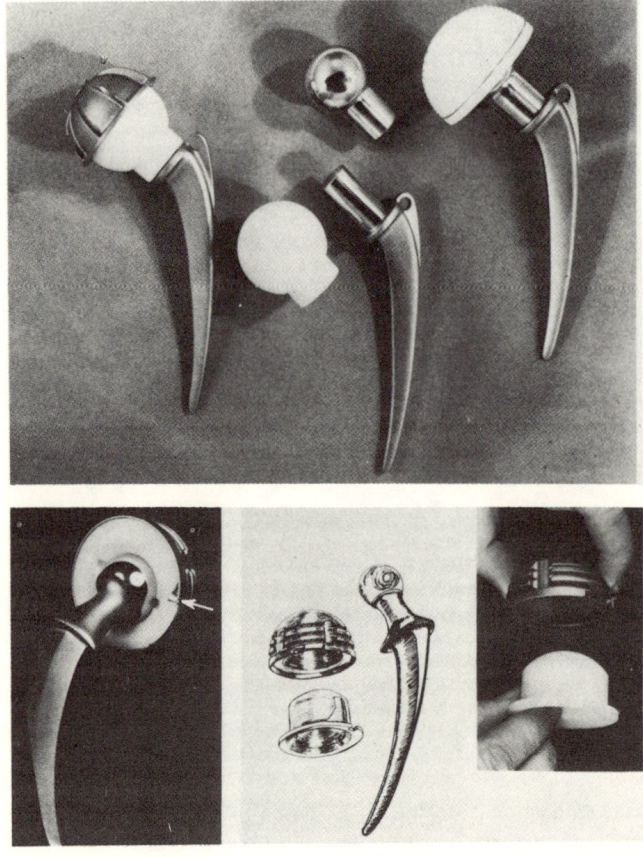

Figure 12-12. The hip prosthesis of Weber and Harris design. (Reproduced from *Clin. Orthoped. Related Res., 72,* 79, 1970, and *81,* 105, 1971, courtesy of J. B. Lippincott Co., Philadelphia.)

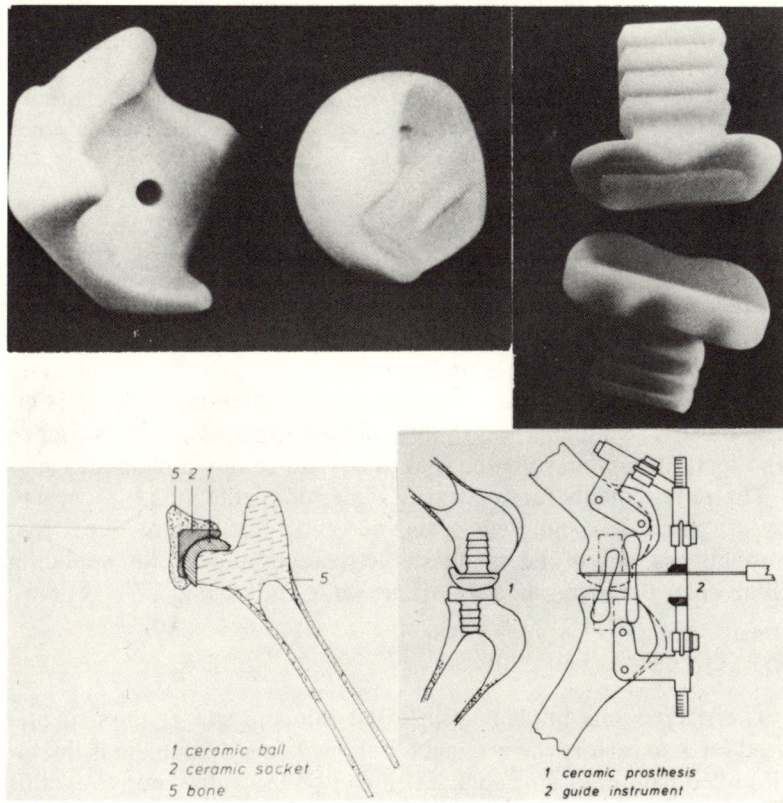

Figure 12-13. Hip (left) and knee joint (right) made from ceramics (alumina). (Reproduced by permission from D. Geduldig, R. Lade, P. Prüssner, H-G. Willert, L. Zichner, and E. Doerre, in *Advances in Artificial Hip and Knee Joint Technology,* pp. 439, 443, ed. M. Schaldach and D. Hohmann, Springer-Verlag, Berlin, 1976.)

toward any environment, including biological, excellent wear resistance, and high compressive strength. Figure 12-13 gives two examples of hip and knee joint prostheses used in animal studies.

Example 12-3

Suggest methods of prosthesis fixation with bone.

Answers

There are four ways of fixation of implants with bone:

a. Direct mechanical fixation using screws, plates, wires, etc., and passive mechanical fixation, such as impaction of the femoral hip stem into the medullary cavity of the femur.

b. Cement, mainly polymethylmethacrylate bone cement.
c. Ingrowth of tissue into the porous or grooved surface of implants.
d. Direct chemical bonding between implant and bone.

The last method is still in experimental stages while (c) is used clinically to some extent. The reader should give some attention to this fixation problem since this is the most difficult to achieve.

12.3. DENTAL IMPLANTS

Implants have been used to support dentures; they can be fixed or removable, partial or full, and sub- or transgingival. The severest challenge is encountered in tooth implants. Like transcutaneous implants, the tooth is exposed to the oral environment, to which no material is completely inert. The added effect of tremendous compressive stress imposed on the teeth during mastication makes the life of the implant short.

The requirements for successful dental implants are (1) biocompatibility, (2) corrosion and wear resistance, (3) high compressive strength and toughness, and (4) adequate viable fixation between the implant and both alveolar bone and mucosal tissue (gingiva, cf. Fig. 7-7).

12.3.1. Endosseous Implants

The endosseous implant is inserted into the site of missing or extracted teeth to restore the original function. The ideal implant is the tooth itself pulled from the same socket if it can be replanted; however, this is not the solution to the problem of tooth replacements. There are many different types of endosseous implant designs, as shown in Figure 12-14. The main idea behind the various root portions of self-tapping screws, spiral, screw-vent, and blade-vent implants is to achieve immediate stabilization as well as long-term viable fixation. The post is covered with an appropriate crown after the implant is fixed firmly (Fig. 12-15).

The difficulty arises when the alveolar bone is resorbed by either too high or too low stress in the normal use of the implant. Also even minute leakage between implant and gingiva will lead to infection and implant failure.

The implant is usually encapsulated by a thin layer of collagenous tissue that is organized in the direction of stress. No mineral deposition close to the implant was found with the screw implant. This indicates that the collagenous tissue acts as a periodontal membrane that distributes and absorbs impact forces of mastication.

The downgrowth of epithelium sometimes is observed around implant posts; it is like that seen in transcutaneous implants. This tends to

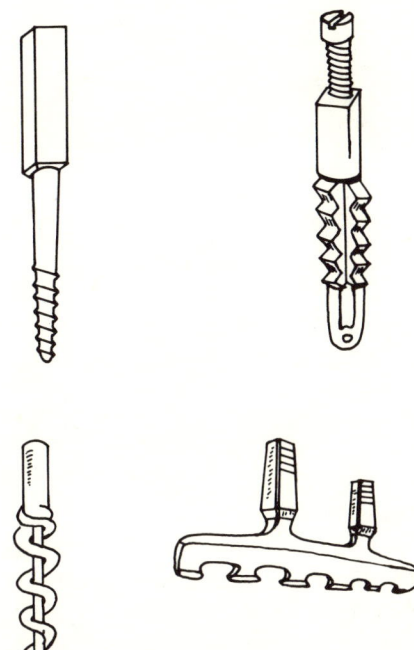

Figure 12-14. Various designs of self-tapping endosseous implants. [Reproduced from D. E. Grenoble and D. Voss, *Biomater. Med. Devices Artif. Organs,* 4(2), 133, 1976, courtesy of Marcel Dekker, New York.]

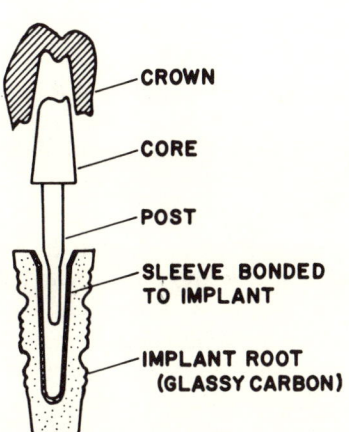

Figure 12-15. Tooth-root-shaped implant fabricated from glassy carbon. [Reproduced from D. E. Grenoble and D. Voss, *Biomater. Med. Devices Artif. Organs,* 4(2), 133, 1976, courtesy of Marcel Dekker, New York.]

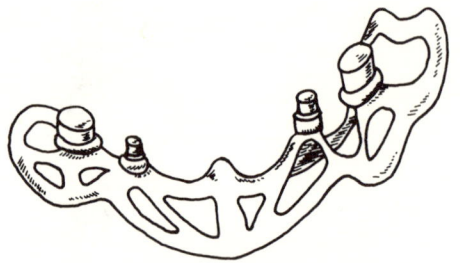

Figure 12-16. A mandibular subperiosteal implant framework cast from cobalt–chromium alloy. [Reproduced from D. E. Grenoble and D. Voss, *Biomater. Med. Devices Artif. Organs,* 4(2), 133, 1976, courtesy of Marcel Dekker, New York.]

support the idea that gingival tissue is similar to skin and that some kind of rejection mechanism is triggered by the presence of a foreign material.

12.3.2. Other Dental Implants

Implants have been successfully used to provide a framework for dentures, as shown in Figure 12-16. The implants can be subgingival or transgingival, although the odds for success are much better with the former (25 years of implantation have been reported) because of the lower chance of infection. Other dental implants include endodontic pin stabilizers, alveolar bone and ridge augmentation, and periodontal defect repair.

12.3.3. Materials Used for Dental Implants

Such metal alloys as cast and wrought cobalt–chromium and Ti-6Al-4V are widely used clinically. Stainless steels are seldom used because the amount of metal used for dental work is much smaller than in orthopedic implants and its cost imposes less restriction. Recently, carbon was used to fabricate endosseous tooth implants for clinical use. Pyrolytic carbon used for percutaneous buttons and artificial heart valve disks is being tested for future use.

Such ceramic materials as dense alumina (Al_2O_3) are attractive for their inertness and high compressive strength if they can be made without any pores. Some ceramic materials, such as calcium phosphate, are being evaluated for filling bone defects, e.g., alveolar and periodontal bone defects. The material is resorbed as the new bone is formed to replace the implant.

Polymethylmethacrylate (PMMA) mixed with inorganic bone (20 w/o) has been tested for bone ingrowth. The main advantage of this material is its ease of fabrication, although its strength is substantially lower than that of bone or other implant materials. Most other materials have been tried in porous form in an effort to form a viable fixation, as shown in Figure 12-17.

HARD TISSUE REPLACEMENT II

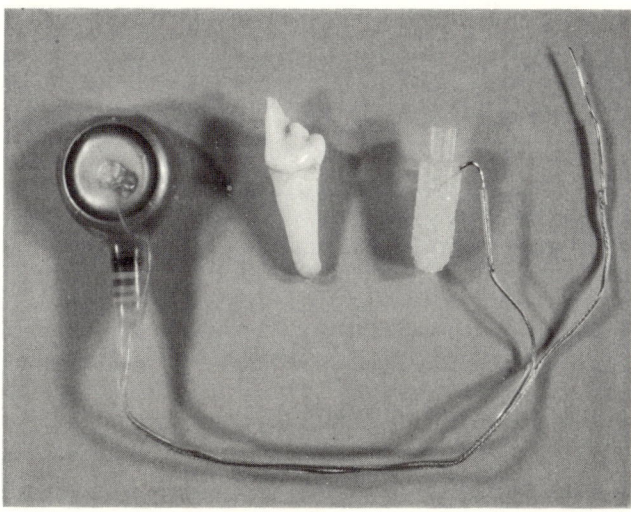

Figure 12-17. A porous dental implant used for electrical stimulation in a tissue ingrowth study. Note the position of the electrode. The implant was placed in the distal root socket of the fourth canine premolar. (From S. O. Young, J. B. Park, G. H. Kenner, B. W. Sauer, B. R. Myers, R. R. Moore, and A. F. Von Recum, Clemson University.)

Example 12-4

During the experiment for tissue growth into the pores of the tooth implant shown in Figure 12-17, a push-out test was performed after sectioning out 2-mm disks (2 for each implant) and one of the curves is given below (the diameter is 4.6 mm).

a. Calculate the maximum interfacial shear stress between the bone and the implant.

b. What is the shear modulus?

c. Is the interfacial strength adequate for fixation of the implant?

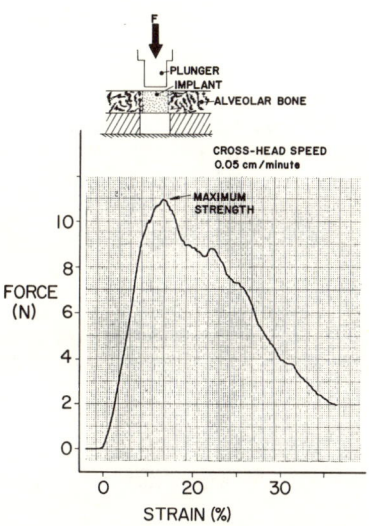

Force–strain curve of a push-out test (the insert shows the test arrangement) of a tooth implant section.

Answers

a. $$\sigma = \frac{F}{A} = \frac{F}{\pi \cdot D \cdot h} = \frac{11 \text{ N}}{28.9 \text{ mm}^2} = \underline{0.38 \text{ MPa}}$$

b. $$E = \frac{\sigma}{\epsilon} = \frac{5.8 \text{N}/28.9 \text{ mm}^2}{0.05} = \underline{4 \text{ MPa}}$$

c. The interfacial strength may not substantially contribute to the total mastication load (because the natural tooth has a conical shape, the applied load is distributed into two major components, shear and compression); therefore, the interfacial shear strength calculated in (a) is a high enough value. (The direct attachment of artificial tooth by tissue growth into the pores of an implant has certain advantages. However, the types of tissues, whether hard or soft, to be grown into to make ankylosis and give the best results have not been established.)

PROBLEMS

12-1. Why are the average maximum values of force at the hip and knee several times the body weight? Propose a method of measuring the forces transmitted through the joints *in vivo*.

Answer
The total energy of a body can be expressed in terms of potential and kinetic energy, i.e.,

$$\text{Total energy} = \text{K.E.} + \text{P.E.}$$
$$\text{K.E.} = \tfrac{1}{2}mv^2$$
$$\text{P.E.} = m \cdot g \cdot h$$

where m is mass, g is gravity, h is height, and v is velocity. If a person is standing still only potential energy will be imposed on the joints because $v = 0$. If a person moves, the kinetic energy will act on the joints and its magnitude will depend on the velocity squared.

12-2. Contrast the integral-hinge-type and mechanical-hinge-type finger joint prostheses.

12-3. List the following materials used to make a joint in the order of least friction:

a. Co–Cr alloy/high-density polyethylene
b. Stainless steel/high-density polyethylene
c. Co–Cr alloy/Co–Cr alloy
d. Al_2O_3/Al_2O_3

Answer
Figure 12-8 shows the order of least friction to the most:
a. S.S./HDPE
b. Co–Cr/HDPE
c. Co–Cr/Co–Cr
d. Al_2O_3/Al_2O_3

12-4. The calcar region of the hip prosthesis frequently undergoes loosening. Some people suggest that this is caused by stress concentration while others claim it results from the lesser stress imposed on the bone tissue, which makes it osteoporotic. How can you reconcile these opposite views?

Answer
The opposite view may result from the fact that the bone constantly undergoes osteogenetic and osteoclastic processes. If the stress is *too small* or *too large*, the osteoclastic process is predominant. Therefore, the implant should be designed to distribute the load according to the body's normal activity before the implantation. Any deviation from this will result in osteoporotic bone.

12-5. Suggest the characteristics of an ideal bone cement for joint prosthesis.

Answer
Because the cement is an implant, all the requirements of implants should be applied to it. However, some additional requirements are unique for the cement:

a. It should be easy to manipulate at the surgical theater.
b. It should not reach a high temperature during setting, which would deteriorate the tissues.
c. It should not creep (which changes dimensions), fatigue, etc.
d. It should be spread evenly and easily around the implant so that no gaps will occur.
e. It should polymerize completely so that no small molecules will be leached out later.

12-6. What biological factors make dental implants so difficult? Suggest the characteristics of an ideal dental implant.

Answer
The percutaneous nature of dental implants make it difficult for them to be successful. The gingiva is in fact a specialized skin that is hard to abut against the implant so that the submerged portion of the implant can be isolated from the oral environment. Another factor is the extremely large compressive stress imposed during mastication, in addition to the cyclic type of loading.

12-7. Give two important characteristics of a material used to construct joints.

Answer
a. The most important factor is wear resistance and such related questions as wear debris.
b. Another important factor in any joint is the friction between the surfaces.

FURTHER READING

J. Charnley, *Acrylic Cement in Orthopaedic Surgery*, Churchill Livingstone, Edinburgh, 1970.
J. Charnley (ed.), "Total Hip Replacement," *Clin. Orthop. Relat. Res.*, 72, 1970.
J. H. Dumbleton and J. Black, *An Introduction to Orthopedic Materials*, chapter 9, Charles C Thomas, Springfield, Ill., 1975.
H. M. Frost, *Orthopaedic Biomechanics*, chapters 18 and 22, Charles C Thomas, Springfield, Ill., 1973.

D. E. Grenoble and D. Voss, "Materials and Designs for Implant Dentistry," *Biomater. Med. Devices Artif. Organs,* 4(2), 133, 1976.

M. Schaldach and D. Hohmann (eds.), *Advances in Artificial Hip and Knee Joint Technology,* Springer-Verlag, Berlin, 1976.

S. A. V. Swanson and M. A. R. Freeman (eds.), *The Scientific Basis of Joint Replacement,* J. Wiley and Sons, New York, 1977.

A. R. Taylor, *Endosseous Dental Implants,* Butterworths, London, 1970.

D. F. Williams and R. Roaf, *Implants in Surgery,* chapters 6 and 8, W. B. Saunders Co., London, 1973.

APPENDIX

SI UNITS

The International System of Units or SI (Le Système International d'Unités) units define the *base units* as follows:

Base Units

Quantity	Unit	Symbol
length	meter	m
mass	kilogram	kg
time	second	s
electric current	ampere	A
temperature	kelvin	K
amount of substance	mole	mol

The *derived units* are as follows:

Derived Units

Quantity	Unit	Symbol	Formula
frequency	hertz	Hz	$1/s$
force	newton	N	$kg \cdot m/s^2$
pressure, stress	pascal	Pa	N/m^2
energy, work, quantity of heat	joule	J	$N \cdot m$
power	watt	W	J/s
absorbed dose	gray	Gy	J/kg

The common prefixes used in this book are as follows:

Multiplication factor	Prefix	Symbol
10^9	giga	G
10^6	mega	M
10^3	kilo	k
10^{-2}	centi	c
10^{-3}	milli	m
10^{-6}	micro	μ

The conversion factors between the various units to SI units are as follows:

To convert from ...	to ...	multiply by ...
angstrom (Å)	m	10^{-10}
inch	m	0.0254
free-fall, standard (g)	m/s²	9.80665
Calorie	J	4.1868
erg	J	10^{-7}
dyne	N	10^{-5}
kg force	N	9.80665
pound (mass)	kg	0.4535924
atmosphere (standard)	Pa	0.1
inch of Hg (60°F)	Pa	3.37685×10^3
lb$_f$/in.² (psi)	Pa	6.894757×10^3
poise	Pa · s	0.1
rad	Gy	0.01

NAME INDEX

Adams, L. M., 155
Agarwal, G., 6
Akahoshi, Y., 143
Alfrey, T., 96
Allgöwer, M., 191
Al-Nakeeb, S., 155
Anliker, M., 107, 120, 129
Argon, A. S., 28
Armeniades, C. D., 184
Azàroff, L. V., 57

Barker, R., 129
Bear, R. S., 101
Bechtol, C. O., 146, 207
Bement, A. L., Jr., 145
Bergel, D. H., 116
Billmeyer, F. W., Jr., 76, 96
Black, J., 129, 207, 231
Blackwood, H. J. J., 129
Bloch, B., 96
Bloom, W., 115
Bokros, J. C., 67, 68, 69
Boretos, J. W., 96
Bourne, G. H., 136, 146
Bowen, H. K., 72
Boyd, S. J., 156
Brighton, C. T., 137
Brooks, C. E., 154
Brown, A., 192
Brown, I. A., 113, 130
Brown, J. H. U., 6
Bruck, S. D., 184
Bulbulian, A. H., 161
Burstein, A. H., 207
Burton, A. C., 117

Chalian, V. A., 159, 161
Charnley, J., 146, 231
Cholvin, N. R., 155
Chvapil, M., 99, 129
Cipolletti, G., 90
Clark, A. E., 70
Cole, J. J., 156
Cooke, F. W., 90, 91, 157
Cooney, D. O., 174, 181, 184
Cottrell, A. H., 28

Daly, C. H., 112, 118, 130
D'arcy, J. C., 192
Dillon, M. L., 150
Doerre, E., 225
Dorman, F., 156
Drane, J. B., 161
Dumbleton, J. H., 3, 207, 231

Eastoe, J. E., 130
Ebner, M. L., 39
Elden, H. R., 129
Ethridge, E. C., 135, 146

Fawcett, D. W., 115
Ferguson, A. B., 143, 146, 207
Flanagan, D., 72
Fleisch, H., 129
Flory, P. J., 96

Fontana, M. G., 57
Forstron, R., 156
Frakes, J. J., 66
Frankel, V. H., 207
Freeman, M. A. R., 190, 216, 232
Friedenberg, Z. B., 137
Fries, C. C., 169
Frost, H. M., 193, 210, 211, 214
Fukaya, H., 120
Fung, Y. C., 107, 120, 129, 231

Geduldig, D., 225
Gendreau, C. L., 199
Ghidoni, J. J., 155
Gibson, P., 118
Gills, L., 149
Gilman, J. J., 72
Gilman, T., 146
Glazener, W. E., 155
Glimcher, M. J., 130
Gomes, M. N., 157
Gorniowsky, M. J., 162
Gould, B. S., 116
Grande, L. A., 118
Green, N. O., 57
Gregor, H. P., 96
Grenoble, D. E., 227, 228, 232
Gurnee, E. F., 96
Gustavson, K. H., 129
Guy, A. G., 57

NAME INDEX

Hall, C. W., 155
Hall, D. A., 129
Ham, A. W., 136
Hardy, J. D., 184
Harkness, R. D., 116, 130
Harris, W. R., 136
Hastings, W. W., 96
Hayden, H. W., 28
Hench, L. L., 70, 135, 141, 146
Hodge, E. S., 143
Hodge, J. W., Jr., 151
Hoffman, A. S., 118, 122, 130
Hohmann, D., 212, 213, 225, 232
Homsy, C., 145, 184
Hoppin, F. G., Jr., 120
Houston, S., 151
Hufnagel, C. A., 157
Hulbert, S. F., 146, 161

Irwin, W. T., 139

Jacobs, J. E., 6

Kadefors, R., 154
Katsura, S., 120
Kenedi, R. M., 118, 130
Kenner, G. H., 140, 199, 229
Kingery, W. D., 65, 72
Knight, J., 156
Kolff, W. J., 176
Konikoff, J. J., 137
Kopp, K., 181
Krane, S. M., 130
Kraus, H., 130
Krock, R. H., 39
Kronenthal, R. L., 96, 185
Kummer, G. H., 107
Kuntscher, G., 207

Lade, R., 225
LaGrange, L. D., 67, 69
Laing, P. G., 143, 146, 207
Laurence, M., 190
Lee, G. C., 120

Lee, H., 96, 161, 175, 176, 185
Leininger, R. I., 185
Leonard, F., 151, 185
Levine, S. N., 146, 161, 162
Lindahl, O., 189
Lyman, D. J., 96

Maibach, H. I., 146, 162
Martin, C. J., 120
Martin, R. L., 154
Martinz, A., 6, 169
Massie, W. K., 2
Matter, P., 191
McClintock, F. A., 28
McEvoy, T., 156
McMahon, J. D., 6, 169
Meares, P., 96
Miller, J., 154
Miller, M. L., 167
Moffatt, W. G., 28, 39, 72
Monson, B., 156
Monypenny, J. H. G., 201
Mooney, V., 155, 162
Moore, R. R., 229
Morgan, J., 150
Morris, L. B., 154
Moyle, D. D., 146, 161
Myers, B. R., 229
Myers, G. H., 185

Neville, K., 96, 161, 175, 176, 185
Newcombe, J. K., 139
Norton, F., 72
Nosé, Y., 156

Oser, Z., 96, 185
Ousterhout, D. K., 151
Owen, M., 129
Ozdemir, D., 157

Parchinski, T., 90, 91
Park, J. B., 118, 122, 129, 140, 199, 229
Parsonnet, V., 185

Paschall, H. A., 70, 141
Paul, J. P., 212, 213
Pauling, L., 30, 39
Peacock, E. E., Jr., 135, 146
Pearsall, G. W., 39, 72
Pearson, P. T., 155
Perren, S. M., 191
Perrone, N., 107, 120, 129
Phillips, R. W., 159
Postlethwait, R. W., 150
Prüssner, P., 225

Ramachandran, G. N., 130
Rawson, R. O., 153
Remington, J. W., 130
Reswick, J. B., 154
Roaf, R., 6, 142, 196, 198, 202, 203, 207, 232
Robert, A. M., 130
Robert, L., 130
Rogers, A., 154
Ross, R., 118, 134, 146
Roth, A. M., 155, 162
Rouse, G., 199
Rovee, D. T., 146, 162
Rüedi, T., 191
Rutter, L. A. G., 149

Sauer, B. W., 229
Scales, J. T., 217
Schaldach, M., 212, 213, 225, 232
Schaube, J. F., 150
Schnittgrund, G. N., 66
Schoen, G. J., 67, 69
Schulz, J. M., 96
Schwartz, S. A., 162
Scribner, B. H., 156
Shrager, M. A., 39, 60
Sivash, K. M., 215
Standish, S. M., 161
Starfield, M. J., 39, 60
Stark, L., 6
Striker, G. E., 156
Stuck, W. C., 207
Swanson, S. A. V., 190, 216, 232

NAME INDEX

Taylor, A. R., 232
Teckhoff, H. A. M., 156
Timoshenko, S. P., 28

Uhlmann, D. R., 72

Van Paasschen, W. H., 156
Van Vlack, L. H., 28, 57, 62, 72
Van Winkle, W., Jr., 135, 146
Vasko, K. A., 153
Venable, C. S., 207
Viidik, A., 130

Von Recum, A. F., 229
Voss, D., 227, 228, 232
Vroman, L., 185

Walker, P. L., 67, 68, 69
Wathen, R., 156
Wesolowski, S. A., 6, 169
Wildnauer, R. H., 113, 130
Wilkes, G. L., 113, 130
Willert, H.-G., 225
Williams, D. F., 6, 142, 196, 198, 202, 293, 207, 232
Wilson, J. N., 217

Winter, G. D., 162
Wulff, J., 28, 39, 72
Wunderlich, B., 80

Yamada, H., 109, 130
Young, A. C., 120
Young, D. H., 28
Young, S. O., 229

Zapffe, C. A., 52
Zichner, L., 225
Ziegler, T. F., 167
Zipkin, I., 130

SUBJECT INDEX

Acetabular cup, 212, 221
Acetabulum, 212
N-Acetylglycosamine, 102
Acrylamide, 166
Acrylic resin, 22, 23
Acrylonitrile, 77
Adenosine diphosphate (ADP), 164
Adhesion, 24, 25
Adhesive
 layer, 25
 surgical, 6, 142, 150
 tissue, 150
Adsorbate, 24
Adsorption, 24
Adventitia, 114, 165
AISI (American Iron and Steel Institute), 201
Alanine, 101
Albumin, 166
Alcohol, 22
Alloy (*see* under specific names)
Alpha S-2, 151
Alumina or aluminum oxide (Al_2O_3), 4, 38, 63, 65, 66, 197, 204, 206, 224, 225, 228
Aluminum, 51, 52, 54, 139, 168, 204
Alveolar bone, 106, 226
Alveolus, 106, 119
Amalgam, 22, 23, 56
Amide (R'CONHR), 74, 98
Amine (RNH_2), 74, 151
Amino acid, 75, 98, 100
Ammonia (NH_3), 37
Amorphous material, 36, 85
Aneurysm, 168

Anionic radical, 166
Ankle, 220
 prosthesis, 222
Anode, 48, 137
Anorganic bone (*see* Bone)
Anticoagulant, 164, 179
Antidiscoloring agent, 74, 140
Antioxident, 74, 140
Aorta, 111, 117, 175
 descending, 175
 human, stress–strain curve, 121
Aortic valve, 171
 wall, 100, 102
Apatite ($Ca_5P_3O_{12}F$), 63, 107
 in bone and teeth, 105
Appositional growth, 136
Arterial prosthesis, 169
Arteriole, 117
Artery, 116, 117, 121, 122, 170
Arthroplasty, mold, 214
ASAIO (American Society for Artificial Internal Organs), 185
Aspartic acid, 101
ASTM (American Society for Testing and Materials), 201, 203, 205
Atomic radius (Table 5-1), 60
Atrioventricular (AV) node, 176
Austenite, 47
Austenitic temperature, 47
 γ phase, 201
Autoclave (*see also* Sterilization, steam)
 of ligament, 117
Autografting, 157, 168
Autoimmune reaction, 133

239

SUBJECT INDEX

Bakelite® (see Polyphenolformaldehyde)
Barium titanate, 61, 62
Base metal, 49
Beeswax, 22, 159
Bending
  of a beam, 18
  moment, 192, 195
    of a bone plate (Fig. 11-2), 190
  stress, 18
Benzalkonium chloride (GBH process), 166
Benzene, 82
Benzoyl peroxide ($C_6H_5COO-OOC_6H_5$), 75
Bernard-Teco assist pump, 175
Bicarbonate, 177
Biocompatibility, 3, 144
Bioglass
  –ceramic interface with bone (Fig. 8-7), 141
  implant, 70
Biological material, 1, 97
Biomaterial
  definition, 1
  surgical uses, 5
Biopotential (see Electrical potential)
Blade-vent implant, 226 (see also Implant, dental)
Blood
  circulation in body (Fig. 10-2), 168
  circulation in the heart (Fig. 10-4), 171
  clotting, 148, 163
  clotting sequence (Fig. 10-1), 164
  coagulation, 163, 164
  compatibility, 163, 164, 165
  effect on strength, 66
  flow, 171
  interface, 168
  plasma, 166
  pressure, 171
Blood vessel walls, 26, 114
  composition changes, 114
  structure (Fig. 7-11), 115
  tension and pressure relationship (Table 7-7), 117
Body-centered cubic, 30, 32
Bonding, 29
Bone, 26, 100, 103, 104, 110, 187
  anorganic, 228
  cement, 90, 142, 145, 199, 213, 218, 226, (also Acrylic cement)
  compact, 106, 124, 135, 197

Bone (cont.)
  demineralized, 109
  fracture healing, 134, 135
  marrow, 106
  properties (Table 7-3), 109
  repair, 187
  resorption, 194
  sequence of fracture healing (Fig. 8-3), 135
  sinusoid, 217
  stress–strain behavior (Fig. 7-8), 107
  structure (Fig. 7-6), 105
Bowman capsule, 177
Branching, 83, 84
Brass, 52
Bronze, 52
Bubble oxygenator, 173
Bundle of His, 176
Burger's vector, 35

Cadmium, 52
Calcar region, 213, 216
Calcium
  aluminate ($CaO \cdot nAl_2O_3$), 4, 65, 66
  ion (Ca^{2+}), 164
  oxide (CaO), 55
  phosphate, 228
  phosphate salt, 106
  titanate ($CaO \cdot TiO_2$), 139
  zirconate ($CaO \cdot ZrO_2$), 139
Callus, 134
Calorie, 22
Calve, 154
Canaliculi, 105
Cancellous bone, 192, 213
Cannula, 153, 155, 156, 157, 167, 180
Capillaries, 117, 131
Carbide, 42, 47, 50
Carbon, 4, 59, 67, 139, 161
  dioxide, 174
  glassy or vitreous, 227
  graphite, 4, 52, 68, 166, 167
  implant, 139
  low temperature isotropic (LTI), 68
  physical properties (Table 5-4), 69
  pyrolytic, 4, 68, 171, 206, 228
  steel, 52
  structure (Fig. 5-8), 67
  tetrachloride (CCl_4), 37
  unassociated, 67

SUBJECT INDEX

Carbonate, 105
Carboxyl group, 98
Carboxylic acid (RCOOH), 74, 151
Carcinogenesis, 142
Cardiac muscle, 111
Cardiovascular system, 5
Cartilage, 103, 111, 135
Cat, 120
Catgut, 139, 148
   breaking strength of (Table 9-1), 149
Catheter, 163
Cathode, 48, 137
Cellophane, 177
Cells, definitions (Table 8-1), 132
Cellulose, 74, 179 (see also Polysaccharide)
Cementing line, 105
Cementum, 106
Chain
   backbone or side, 82
   end-to-end distance, 81
   extended, 80
   fold, 80
   mobility, 79
Charcoal, 179
Charnley hip prosthesis, 217, 221
Chemisorption, 24
Chloride (Cl), 177
Chondroblast, 132, 134
Chondroitin, 75, 102
   structure (Fig. 7-4), 103
   sulfate, 75, 102, 103, 112, 165
Chromic salt, 148
Chromium (Cr), 32, 51, 52, 143, 202
Citrate, 105, 164
Clearance, curve for dialysate versus flow rate, 180
Close-packed, 32
Cobalt (Co), 143
Cobalt-60, 89
Cobalt-base alloy, 47
Cobalt-chromium alloy, 21, 36, 171, 197, 200, 202, 204, 206, 217, 221, 224, 228
   cast, 203
   composition (Table 11-5), 203
   mechanical properties (Table 11-6), 203
Cohesion, 25
Cold-working, 46, 55, 202
Collagen, 5, 75, 98, 104, 106, 108, 112, 116
   amino acid content (Table 7-1), 99
   fibers, 111

Collagen (cont.)
   fibrils, 106
   matrix in bone and teeth, 106
   reconstituted, 5, 90, 157
   sutures, 148
   synthesis after wound, 135
   type I, II, III, 100
Collagenase, 118, 123, 132, 133
Composites, 4
Compression, 7
   dynamic plate, 191, 206
Compressive
   strain, 193
   stress, 209
Connective tissue, 133
   structure–property representation (Fig. 7-15), 118
Contact angle, 25, 165
   of materials (Table 2-4), 25
Conversion factors, SI units, 234
Coordination number, 32, 60
Copolymer
   polyethylene glycol and polyethyleneterephthalate, 180
   polymethylmethacrylate and polystyrene, 145, 219
   vinyl chloride and acetate, 157, 159
Copolymerization, 180
Copper, 10, 22, 24, 42, 46, 54, 56, 165
   deficiency, 100
   ion (Cu^{2+}), 100
   steel, 52
Cornea, 102, 103
Corrosion
   cell, 48
   fatigue, 50
   intergranular, 202
   iron-hydrogen cell (Fig. 4-7), 49
   particles, 142
   rate, 53
   resistance, 200, 204
   stress, 50
Cosmetic implant, 5
Cotton, 148, 150
Covalent bonding, 29, 37
Cracks, effect on strength (Fig. 5-5), 63
Crazing, 91
Creatinine, 177, 180
Creep
   recovery, 18
   testing, 126

Cross-linking, 67, 81, 83, 148, 166, 180
Cross-sectional area, 9, 54
Crystal system, 30, 31
Crystalline, 31
Crystallinity, 86
Crystallization, 80, 85
Cuprophane®, 180
Cutter heart valve, 172
(Alkyl-α-) Cyanoacrylate, 151
(Ethyl-2-) Cyanoacrylate, 151
(Methyl-2-) Cyanoacrylate, 151, 157
Cycles, 13
Cysteine, 100

Dacron®, 4, 161, 170 (*see also* Polyester; polyethyleneterephthalate)
Dashpot, 14
De Bakey left ventricular by-pass, 175
Deciduous teeth, 106
Defect, 34
Density, 108, 109, 204, 205
Dental bridge, 205
  implant, 226
Dentin, 22, 106
Dentinal tubules, 106
Denture, 228
Depolymerization, 88
Dermatan sulfate, 103
Dermis, 112
Desmosine, 100
  structure (Fig. 7-3), 101
Deterioration
  of ceramics, 65
  of polymers, 88
  in vivo, 89
Dialysance, 180
Dialysate, 177
Dialysis, home, 180
Diamond (C), 29, 63, 69
Diffusional process, 210
Digestive system, 5
Dimethyl siloxane, 84
Dimethyldichlorosilane, 84
Dipole, 30, 62
Dislocation, 35, 46, 47, 79
Disk type heart valve, 172
Dog, 153, 154, 155, 158, 167, 169
Dynamic load, 188, 195
Dyne, 24

Ear, 112
Eastman 910, 151
Elastase, 118
Elastic
  deformation, 9
  fibers, 111
  force, 194
  region, 8
Elastin, 75, 98, 100, 116, 117
  composition, 102
  variation in arterial wall (Fig. 7-13), 116
Elbow, 210
  prosthesis, 220, 222
Electrete, 166
Electric potential, 136, 187
  of fractured rabbit tibia (Fig. 8-5), 137
Electrical
  charge, 30
  dipole, 61
  stimulation, 134, 137, 153, 229
Electrode, 184
  potential, 48, 56
  of various ions (Table 4-2), 50
Electrolyte, concentration of, 49
Electronegativity, 29, 30, 59
Electrons, sea of, 41
Emboli, 148, 163
Enamel, tooth, 22, 106
Encapsulation, 133, 139, 147
Endocardium, 184
Endogenous heat, 153
Endosseous implant, 227 (*see also* Implant, dental, Fig. 12-14)
Endosteum, 105
Endothelial cells, 131, 133
Endurance limit, 13, 50
Energy, bond, 22
Engineering stress-strain curve, 9
Enzyme, 75, 142, 163
  fibrinolytic, 133
Epidermis, 112
Epiphysis, 136
Epithelial layer cell, 153
Epithelium, 152, 226
Erg, 24
Erythrocyte, 131, 132
Ethane, 68
Ethylene ($CH_2=CH_2$), 37, 75
Ethylene oxide, 89
Eutectic, 44

SUBJECT INDEX

Eutectoid, 44
Exothermic polymerization, 217
Expansion, coefficient of, 23
Extracorporeal, 6, 163
Extraoral prosthesis, requirements, 159
Exudate, 131, 133
Eye, 102

Face-centered cubic, 30, 31
Facia, 114
Failure strength, 9
Fat embolism, 217
Fatigue, 13
FDA (Food and Drug Administration), 159
Felt, Dacron®, 153, 154, 155, 156
Femoral
 artery, 116
 head, 212
 osteotomy, 195
 plate, 192
Femur, 109, 145, 190, 192, 197
Ferrite (α phase steel), 47, 50, 51, 202
Fibrin, 133, 164, 169
Fibrinogen, 132, 133
Fibroblast, 132, 133, 169
Fibula, 190
Filler, 144, 151
Film oxygenator, 173
Filter, disposable, 180
Filtration, membrane, 219
Finger (joint prosthesis), 220, 223 (see also Implant, finger)
Flat plate artificial kidney, 179
Flow rate, 180
Fluidized bed, 68
Fluoride, 105
Fluorine (F), 73
Force
 applied during walking, 212, 213
 equilibrium, 25
 on hip and knee joints (Table 12-1), 212
Foreign body giant cell, 138
Formed blood elements, 163, 167, 171
Fracture
 bone, 134
 fixation, internal, 188
 plate (Fig. 11-1), 189, 190
Free radical, 75, 76
Free volume, 36, 37

Freeman–Swanson total knee, 221
Freezing temperature, 22
Friction, coefficient of, 112
Frictional torque versus applied load for hip prosthesis (Fig. 12-8), 217
Fringed micelle, 80

Galvanic corrosion, 50
 series (Table 4-3), 52
 summary of, 51
Gel, viscoelastic, 102
Gelatin, 166
Genitourinary system, 5
Giant cell, 132, 138
Gingiva (gum), 106, 226
Glass
 ceramic, 70
 direct bonding to bone, 140
 fibers, 64
 surface, 165, 167
 transition temperature (T_g), 37, 82, 87
Glomerulus, 177
Glucose, 74, 177
Glutamic acid, 101
Gluteraldehyde, 166
Glycerin, 22
Glycine, 98, 101
Glycoprotein, 107
Glycosaminoglycan, 102
Goat, 155
Gold (Au), 22, 24, 168
Grain, 35, 50, 51, 203
 boundary, 35, 49, 50, 51
Granular tissue, 138
Granulocytes, 132, 133
Graphite (see Carbon)
Greater trochanter, 21
Ground substance, 103, 112
Growth plate, 136
Guinea pig, 153
Gy, 89

Harrington spinal distraction rod, 198
Harris design hip prosthesis, 224
Haversian system, 105
Healing pattern of arterial prosthesis (Fig. 10-3), 169
Heart, artificial (Fig. 10-8), 176

Heart assist devices (Fig. 10-7), 172, 175
Heart/lung machine, 165
Heart valves, aortic, 111, 171
  various types (Fig. 10-5), 172
Heat of fusion, 22
Helium (He), 174
Helix, 98
  structure (Fig. 7-1), 99
Hemolysis, 163, 167, 171
Heparin, 164, 165, 175, 180, 183
  sulfate, 165
Heparinization, 166
Hexagonal structure, 31, 32
  lattice, 67
Hip
  joint, 210
  joint force during walking (Table 12-1, Fig. 12-3), 212, 213
  joint loading (Fig. 12-1), 210
  joint replacement, 210
  nail (see Nail)
  prosthesis, 215
Histidine, 102
Histiocyte, 133
Hollow fiber, 178, 183
Homografting, 157
Hooke's law, 8, 20
Hoop stress, 218
Hormone, 134
Human, 154, 156
Humerus, 109
Hyarulonic acid, 75, 102
  structure (Fig. 7-4), 103
Hyarulonidase, 102, 103
Hydrocarbon, 68
Hydrogel, 166
Hydrogen, 51
  electrode, 49
  peroxide (H_2O_2), 37
Hydrolysis, 91
Hydrophilic, 165
  polymers, 90
  segment, 180
Hydrophobic, 100, 165
  polymers, 90
  segment, 180
Hydroquinone, 219
Hydroxyapatite, 108 (see also Apatite)
Hydroxyethylmethacrylate (poly-HEMA), 166

Hydroxyl ion (OH^-), 89, 105
Hydroxyproline, 98
Hypochloride, 89
Hypro activity, specific, 134

Immunogenic agent, 133
Impact strength, 12
Imperfection, structural, 46
Implant
  ankle, 220, 222
  biodegradable, 142
  breast, 158, 159
  cellular response, 138
  elbow, 220, 222
  endosseous dental (tooth), 226
  finger, 223
  heart valve, 170
  hip joint, 215, 216
  knee joint, 221
  mandibular, subperiosteal, 228
  materials (Table 1-1), 4
  maxillofacial, 159
  percutaneous (Table 9-2), 152, 153
  permanent, 5
  reconstructive, 158
  replaceability, 217
  requirements, 3
  shoulder, 220, 222
  skin, 157
  stability, 188
  tissue response (Fig. 8-8), 131, 142
  transient, 6
  vascular, 168
  vein, 168
Implantation, effect on polymers (Table 6-5), 91
Impurity, 34
Incompetence, 171
Inconel, 52
Infection, 34, 149, 193
Inflammatory reaction, 131
  acute, 131
  chronic, 133
Initiator, 75
Intermetallic compound, 41
Interstitial solid solution, 42
Interstitial system of bone, 105
Intertrochanteric osteotomy, 193
Intima, 114, 165

SUBJECT INDEX

Intraaortic balloon, 175
Intramedullary
  cavity, 194, 217, 220
  device, 194
  rod, 56, 70
Iodine (I_2), 30
Ion, released, 53
Ionic
  bonding, 29, 37
  fluid, 97
  radius (Table 5-1), 60
Iron (Fe), 29, 38, 51, 52, 54, 143
  oxide (FeO), 55
Irradiation, 134
Irridium (Ir), 205
Isodesmosine, 100
  structure (Fig. 7-3), 101
Isoprene rubber, 81
Isotope, 89
Ivalon®, 158 (see also Polyvinyl, alcohol)

Jarvik-III-type artificial heart, 176
Joint
  arthritic, 220
  articulation, 209
  cartilage, 112
  force (Table 12-1), 212
Joules (J), 22
Judet femoral head, 220

Kalke–Lillehei heart valve, 172
Kelvin model (see Voigt model)
Kidney, 143, 151, 176
  diagram (Fig. 10-10), 178
  dialyzer (Fig. 10-11), 178
  machine, 175, 177
  portable or wearable, 180
Kirschner wire, 188
Knee, 210
  joint force during walking (Table 12-1, Fig. 12-3), 212, 213
  joint implant, 220
  joint structure (Fig. 12-1), 210
  prosthesis, various designs (Fig. 12-9), 221
Knot, surgical, 148
Krebs–Ringer solution, 123
Kyphosis, 198

Lacuna, 105
Langer's line, 112
Laplace equation, 117
Lattice spacing, 30
Lead, 24, 52, 54
Leukocytes, 131, 132
Lever rule, 43
Ligament, 100, 117, 210
Ligamentum nuchae, bovine, 117, 118, 119
Line defects, 35
Lipid adsorption, 91, 172
Liquidus line, 43
Liver, 143, 151
Loosening of implant, 2
Lordosis, 198
Lost wax investment, 204
Lung, 103, 143
  mechanical and chemical characteristics (Table 10-2), 174
  walls, 119
  stress–strain behavior (Fig. 7-16), 120
  volume–pressure curve (Fig. 7-17), 120
Lymphatics, 132
Lysine, 99, 102
Lysinonorleucine, 100
  structure (Fig. 7-3), 101
Lysyl oxidase, 100

Macrophage, 132, 133, 138
Magnesium, 52
Mandibular
  bone, 106
  subperiosteal implant, 228
Manganese, 204
Marrow cavity, 105, 135, 195
Martensite, 47
Mastectomy, 160
Mastication, 209, 226
Maxillary bone, 106
Maxillofacial
  augmentation, 158
  implant, 147
  prosthetics, 159
Maxwell model, 15
McKee hip prosthesis, 217
McKeever tibial plateau, 221
Mechanical properties, 7
Media, 114
Melting temperature, 22

Membrane, 163, 177
  coil, 177
  oxygenator, 173
  silicone rubber, permeability (Table 10-3), 174
Mercury
  battery, 22, 24, 25
Mesenchymal cells, 132, 133
Metallic bonding, 29
Metals, 4, 41, 171, 200
Methane, 68
Methyl acrylate (CH_2=CH−$COOCH_3$), 77
Methyl methacrylate, 77, 92
Methylchloride (CH_3Cl), 84
Methylene iodine (CH_2I_2), 25
Microfibrils, elastin, 102, 111
Mineral deposition, 226
Mineralized tissues, 104
Miniscus, 211
Mitral valve, 170
Modulus
  ceramics (Table 5-3), 65
  of elasticity, 9, 74, 82, 125, 195
  Young's, 9
Moh's hardness scale, 63
Molecular weight, 73
  number average, 78
  of tropocollagen, 100
  weight average, 78
Molybdenum (Mo), 52, 143, 202, 203, 204
Monel metal, 52
Monomer, 75, 140, 214
Mononuclear cells, 132, 133
Mucopolysaccharide, 102, 112, 165
  acid, 100
  −protein complexes, 103
Multinuclear giant cells, 133
Muscle, 143, 210
Muscular skeleton system, 5

Nail, hip, 194, 195
  cross-section (Fig. 11-8), 196
Natural rubber, 86 (see also Poly(cis-)-isoprene)
Necking, 10
Negative electric charge, 166
Neointima, 169, 170
Nephron, 177
  diagram of (Fig. 10-10), 178
Nervous system, 5

Neutral axis (NA), 18
Newtons (N), 7
Nickel (Ni), 42, 46, 52, 201
Niobium (Nb), 143, 202
Nitrogen (N), 174
Noble metal, 49
Norepinephrine, 122
Nose tip, 112
Nylon, 4, 79, 139, 142, 148, 149, 158, 168, 169, 177 (see also Polyamide)

Onlay fracture plate, 188
Organs, artificial, 175
Orientation of chains, 85
Orlon®, 150 (see also Polyacrylonitrile)
Orthopedic fixation devices, 6
Osmium (Os), 205
Osteoblast, 132, 134
Osteoclastic activity, 187
Osteogenesis, 70
Osteogenic, 187
  activity, 187
  cells, 134
Osteon, 105
Osteotomy
  devices (Fig. 11-7), 196
  femoral, 195
Oxalate, 164, 174
Oxygen (O_2), 51
  concentration, 49
  tension, 134
Oxygenator, types (Fig. 10-6), 164, 172, 175
Ozone (O_3), 37

Pacemaker, 163, 175, 176, 177, 184
  nuclear-powered, 176
  typical (Fig. 10-9), 177
  tip, 205
Palladium (Pd), 205
Paraffin, 22, 25, 159
Pascal (Pa), 7
Passivation, 52
Passivity, 51
Pearlite, 50
Pendulum, 12
Penis, artificial, 160
Peptide, 75, 98
Percutaneous device, 152

SUBJECT INDEX

Periodic table, 30
Periodontal
  bone defect, 228
  ligament (or membrane), 106, 226
Periosteum, 105, 134
Permanent teeth, 106
Peroxide, 219
pH, 132
Phagocyte, 132
Phase, 41
  boundaries, 24
  diagram, 41, 43
    Cu–Ni (Fig. 4-3), 43
    Cu–Ag (Fig. 4-4), 44
    Fe–C (Fig. 4-5), 45
    Ag–Sn, 57
  separation, 42
  transformation, 22
Phenolic solution, 89
Piezoelectric phenomenon, 61, 187
Pig, 155, 169
Pitting corrosion, 51
Plasma, 131, 166
  protein, 180
Plaster of Paris, 65
Plastic
  deformation, 9, 64
  region, 8
Plasticizer, 74, 82, 90, 91, 140, 151
Plasticizing, 89
Platelet, 131, 132, 163, 164
Platinum (Pt), 22, 49, 53, 56, 205
Platinum–10% Irridium alloy, 176
Plexiglass® (*see* Polymethylmethacrylate)
Point defect, 34
Poisson's ratio (ν), 26, 54
Polar molecule, 30
Polyacetal, 89
Polyacrylonitrile, 149, 169
Polyamide, 76, 79, 86, 89, 90, 91, 161 (*see also* Nylon)
Polycarbonate (bisphenol A), 87
Polychlorotrifluoroethylene, 171
Polycrystalline material, 35
Polydimethylsiloxane (*see* Silicone, rubber)
Polydispersity, 78
Polyester, 76, 149, 161, 168, 169 (*see also* Polyethyleneterephthalate)
Polyethylene, 3, 26, 71, 73, 79, 83, 86, 87, 89, 91, 92, 148, 161, 165, 166, 206, 217, 221, 224

Polyethylene (*cont.*)
  oxide, 92
  unit cell structure, 81
Polyethyleneterephthalate (polyester), 82, 86, 91, 139
Polyglycine, 98
Polyglycolic acid (PGA), 148, 149
Poly(cis-)isoprene (natural rubber), 84, 88
Polymerization, 74
  addition or free radical, 75, 84, 88
  condensation, 74
  degree of, 77
Polymers, 4, 43
  advantages, disadvantages as implants, 73
  condensation, 76
  deterioration, 88
  glassy, 87
  linear, 79
  mechanical properties (Table 6-4), 86
  natural, 74
  vinyl, 75
Polymethylmethacrylate (PMMA), 26, 28, 86, 87, 89, 90, 91, 95, 142, 159, 161, 165, 168, 197, 206, 218, 228
Polymorph, 150
Polymorphonuclear cells, 131
Polyolefin (*see* Polyethylene and Polypropylene)
Polyoxymethylene (POM), 171
Polypeptide, 75
Polyphenolformaldehyde, 80
Polypropylene, 78, 86, 90, 91, 148, 149, 169, 171, 220
Polysaccharide, 74, 102, 210
Polysiloxane (*see* Silicone, rubber)
Polystyrene, 81, 92
Polytetrafluorethylene (PTFE), 73, 86, 90, 91, 139, 140, 142, 167, 168, 169, 170, 173, 220 (*see also* Teflon®)
Polyurea, 76, 161
Polyurethane, 76, 161, 166
  rubber, 159
Polyvinyl
  acetate, 88
  alcohol, 88, 90, 158, 159
  chloride, 86, 88, 89, 91, 92
Polyvinylidenechloride, 92
Poppet, heart valve, 171
Porcelain, 22
Pores, 64
  effect on strength (Fig. 5-6), 65

Porosity, 65
Porous
  dental implant, 229
  electrode, 176
  vascular prosthesis, 168
Potassium (K), 177
Power consumption, 175
Precipitation-hardening, 44, 47, 55, 202
Preclotting, 169
Preferred orientation, 68
Primary teeth (see Deciduous teeth)
Primate, 153
Probability of failure, 2
Procollagen, 134
Proline, 98, 101
Propagation of polymer chains, 77
Propane, 68
Propyl orthosilicate($Si(OCH_2CH_2CH_3)_4$), 85
Propylene oxide, 89
Protein, 74, 98, 163
  structure (Fig. 7-1), 99
Proteoglycan, 102
Proteolytic enzyme, 132
Pseudoendothelial layer, 148
Pseudointima, 169
Pulp, tooth, 106
Pus, 131
Pyrex® glass, 65
Pyrolidine, 100
Pyrolysis, 68

Quartz (SiO_2), 25, 61, 63
Quasiequilibrium, 36
Quaternary salt, 166
Quenching, 47

Rabbit, 136, 137, 143, 156, 157
Radiation, γ, 219
Radius, 109
  atomic and ionic (Table 5-1), 60
  and coordination number (Fig. 3-8), 34
  of an ion, 60
  ratio, 34, 59
Random chain scission, 88
Rat skin, 135
Rayon, 155 (see also Polyacrylonitrile)
Reconstituted material, 5

Recrystallization temperature, 46
Red blood cell, 163
Regurgitation, 171
Relaxation time, 16, 27, 95
Reliability, 2
Repair, cellular response to, 133
Repeat unit, 73
Respiratory system, 5
Retardation time (λ), 18
Rhodium (Rh), 205
Rib, 112
Room temperature vulcanizing silicone rubber (RTV), 84
Ruthenium (Ru), 205

Safety factor, 189
Salt (NaCl), 29, 37
Scar, 133
  tissue formation, 150, 176
Schulz design finger prosthesis, 220, 223
Schumacker–Burns electrohydraulic heart, 176
Screw, 188
  pull-out strength (Table 11-1), 189
Screw-vent implant, 226 (see also Implant, dental)
Section moduli, 20, 21, 195
Self-tapping screw implant (see also Implant, dental)
  various designs (Fig. 12-14), 227
Semiconductivity, 30
Semicrystalline polymers, 87
Sewing ring, valve, 171
Shear, 7, 63
  modulus, 94
Sheep, 148, 153, 156
Shock absorber, 14
Shoulder joint, 220
  prosthesis, 222
SI unit, 233
Side
  chain substitution, 83
  group, 98
Silastic®, 4, 144, 153, 154, 155, 156
Silica, 61, 65, 84, 133, 144
Silicate, 61, 62
Silicon (Si), 204
Silicone
  fluid, 158, 159
  rubber, 73, 84, 86, 91, 142, 158, 159,

SUBJECT INDEX

Silicone (*cont.*)
    161, 166, 170, 171, 173, 174, 176, 220
Silk, 139, 148, 150
    fibroin, 76
Siloxane, 62
Silver (Ag), 22, 52, 56, 168
Simple cubic, 30
Simplex-P bone cement, 218
Single-needle dialysis, 181
Sinoatrial (SA) node, 175
Sivash prosthesis (hip joint), 215, 217, 224
Skin, 100, 102, 103, 111, 112, 147, 160
    artificial, 152, 157
    canine, 114
    graft, 150
    implant, 152
    microstructure (Fig. 7-10), 113
    rabbit, 137
    stress–strain curve, human (Fig. 7-9), 112
Skull, 109
Slip of ionic and nonionic bond materials (Fig. 5-4), 63
Smooth muscle, 114, 122
Soda-lime glass, 25
Sodium, 177
Soft tissues
    healing, 134
    replacement requirements, 147
Sol, viscous, 102
Solidus line, 43
Solubility limit, 42
Solution-perfused surface, 165, 166
Space lattice, 31
Specific
    heat, 22
    strength, 85
Spinal fixation, 198
Spiral implant, 226 (*see* Implant, dental)
Spleen, 143, 151
Spongy bone, 105, 109, 134
Spring, 14
St. George design finger joint prosthesis, 223
Stanmore hip prosthesis, 217
Starr-Edwards heart valve, 172
Static loading, 65
Steel, 52, 65, 161, 165, 200, 217, 221
    316, 316L Stainless, 4, 53, 54, 197, 201, 204, 206
    316 specification (Table 11-3), 201

Steel (*cont.*)
    mechanical properties (Table 11-4), 202
    partial phase diagram (Fig. 11-10), 201
    plain, 54
    suture, 148, 149
Steinman pin, 188, 199
Stellite®, 202 (*see also* Cobalt-chromium alloy)
Steric rigidity, 100
Sterilization
    chemical, 89
    dry, 88
    radiation, 89
    steam (autoclave), 89
Sternum, 112
Strain, 7, 54
    rate, 14
    -hardening, 46
Streaming potential, 165
Strengthening mechanism, 46
Stress, 7
    concentration, 63, 64
    relaxation, 16, 107, 122, 127
Stress–strain curves
    alveolar wall (Fig. 7-16), 120
    aortic wall (Prob. 7-2), 121
    bone (Fig. 7-8), 107
    ligamentum nuchae (Fig. 7-14), 118
    lung wall (Fig. 7-17), 120
    skin (Fig. 7-9), 112
Stretch ratio, 8
Structure of solid, 29
Structure–property relationship of tissues, 104
Styrene ($CH_2=CH-C_6H_5$), 77
Substitutional solid solution, 42, 46
Sulfur (S), 84
Supercooling, 36, 87
Supracondylar fracture, 192
Surface
    charge, 165
    energy (tension), 24
    nonthrombogenic, 165
    property, 24
    roughness, 165
    tension of materials (Table 2-2), 24
    wettability, 165
Suture, 6, 142, 148
    properties *in vivo* (Table 8-1), 139
    strength, 151
Swanson design finger joint prosthesis, 223

Synovial fluid, 102, 210
Systemic effect by implant, 142

Talc ($Mg_3Si_4O_{10}(OH)_2$), 63
Tantalum (Ta), 159, 205
Tapes, surgical, 149
Teflon®, 4, 174 (*see also*
    Polytetrafluoroethylene)
Tempering, 47
Tendon, 103, 110, 111, 114, 210
Tensile
  force, 110
  strength (ultimate), 9
  strength of skin wound (Fig. 8-2), 135
Tension, 7
  wall, 117
Termination of polymerication, 77
  disproportionate chain, 77
Testicles, artificial, 160
Tetrahedral chains, 62
Thermal conductivity, 23, 109
  of materials (Table 2-1), 22
Thermal properties, 22
  enamel and dentin (Table 7-4), 109
  of polymers, 87
Thermoplastic polymer, 84
Thermosetting polymer, 84
Thigh, loading elements of (Fig. 12-2), 211
Thrombin, 164
Thrombogenicity of various surfaces (Table 10-1), 167
Thromboplastin, 164
Thromboresistance, 165
Thrombus, 163, 170
Tibia, 109
  rabbit, 137
Tibial plate, 192
Tie line, 43
Tin (Sn), 52, 57, 204
Tissue growth, 153
  ingrowth, 169, 171, 176
Titanium (Ti), 47, 51, 139, 143, 154, 171, 200, 202
  alloys (usually Ti-6Al-4V), 4, 204, 206, 224, 228
  mechanical properties (Table 11-7), 205
  oxides (TiO_2), 4, 139, 204
Toluidine, N,N-dimethyl-p-, 219
Tooth, 26, 104, 106
  implant, 226

Tooth (*cont.*)
  physical properties (Table 7-4), 109
  root, 106, 229
  socket, 106
  structure (Fig. 7-7), 106
Toughness, 11
Trabeculae, 106, 134
Trachea, 112
Transcutaneous (*see* Percutaneous device)
Transformation process, 47
Transition element, 30
Transplantation, 5
Tridodecylmethylammoniumchloride (TDMAC), 166
Tropocollagen, 99, 100
Tropoelastin, 102
True stress–strain curves, 9
Trunk muscle, 198
Tube, 163
Tumor, 140
Tungsten, 54
Twin coil artificial kidney, 179

Ulna, 109
Ultrafiltration, 179
Ultraviolet light, 76
Umbilical cord, 102, 103
Urea, 177, 180
  nitrogen, 181
Uronic acid, 177

Vacant lattice, 34, 35
Vagina, artificial, 160
Valence electrons, 29, 41
Valine, 101
Van der Waals force, 30
Vandium (V), 47, 204
Vascular implant, 163, 165, 168
Vasodilatation, 131
Vein, 117, 156, 168, 177
Velour, 154, 155, 157
Vena cava, 117
Ventricle, 170, 184
Venules, 117
Vertebra, 109
Viability, 97
Vinertia®, 202 (*see also* Cobalt–chromium alloy)
Vinyl acetate ($CH_3COOCH=CH_2$), 77

Vinyl chloride ($CH_2=CHCl$), 77
Vinylidene chloride ($CH_2=CCl_2$), 77
Viscoelastic polymer, 213
Viscoelasticity, 13
Viscosity, 14, 94
Visking®, 180
Vitallium®, 4, 47, 143, 145, 159, 202 (see also Cobalt-chromium alloy)
Voigt model, 15, 93, 94, 124
Volkman's canal, 105
Vulcanization, 81, 84

Waldius total knee, 221
Water, 22, 25, 37, 66
Watt (W), 23, 233
Wax, paraffin, bee, 159
Wear, 2, 220
Weber design hip prosthesis, 224 (see also Implant, hip joint)
Weigert's resorcin-fuchsin, 101
Welding, 25
Wetting, 25
White blood cell, 163
Wire, 188
Wolff's law, 107
Work-hardening, 46, 203
Wound healing, 131
    sequence (Fig. 8-1), 134
    strength, 134, 135

Yield point, 8
    offset, 9

Zeta potential (see Streaming potential)
Zinc, 24, 52
Zirconium oxide (ZrO_2), 139